FORENSIC PLANT SCIENCE

FORENSIC PLANT SCIENCE

JANE H. BOCK

DAVID O. NORRIS

ELSEVIER

AMSTERDAM • BOSTON • HEIDELBERG • LONDON
NEW YORK • OXFORD • PARIS • SAN DIEGO
SAN FRANCISCO • SINGAPORE • SYDNEY • TOKYO
Academic Press is an imprint of Elsevier

Academic Press is an imprint of Elsevier
125 London Wall, London EC2Y 5AS, UK
525 B Street, Suite 1800, San Diego, CA 92101-4495, USA
225 Wyman Street, Waltham, MA 02451, USA
The Boulevard, Langford Lane, Kidlington, Oxford OX5 1GB, UK

Notices
Knowledge and best practice in this field are constantly changing. As new research and experience broaden our understanding, changes in research methods, professional practices, or medical treatment may become necessary.

Practitioners and researchers must always rely on their own experience and knowledge in evaluating and using any information, methods, compounds, or experiments described herein. In using such information or methods they should be mindful of their own safety and the safety of others, including parties for whom they have a professional responsibility.

To the fullest extent of the law, neither the Publisher nor the authors, contributors, or editors, assume any liability for any injury and/or damage to persons or property as a matter of products liability, negligence or otherwise, or from any use or operation of any methods, products, instructions, or ideas contained in the material herein.

ISBN: 978-0-12-801475-2

British Library Cataloguing-in-Publication Data
A catalogue record for this book is available from the British Library

Library of Congress Cataloging-in-Publication Data
A catalog record for this book is available from the Library of Congress

Publisher: Shirley Decker-Lucke
Acquisition Editor: Elizabeth Brown
Editorial Project Manager: Joslyn Chaiprasert-Paguio
Production Project Manager: Lisa Jones
Designer: Maria Ines Cruz

Typeset by TNQ Books and Journals
www.tnq.co.in

Dedication

The many to whom we must express gratitude and to whom we dedicate this book fall into two categories: the knowledgeable plant scientists and those in the forensic science community who work for justice.

Contents

5. Cases Using Evidence from Plant Anatomy

6. Forensic Plant Taxonomy

7. Plant Taxonomy Cases

8. Plant Ecology

9. Forensic Plant Ecology Cases

10. Additional Approaches in Forensic Plant Science

11. Summation and a Look to the Future

Please visit: http://booksite.elsevier.com/9780128014752 for supplementary materials.

Author Biographies

Jane H. Bock, PhD

Dr Bock is a professor emerita in biology at the University of Colorado, Boulder. She received her bachelor's degree from Duke University, master's degree from Indiana University and PhD (1966) from the University of California at Berkeley. All her degrees are in Botany. She taught, carried out research, and published scientific work in population ecology and forensic botany at Boulder for over 30 years. Officially retired from teaching, she continues to do research as a forensic botanist and serves as an expert witness for the defense or the prosecution in homicide cases. She also lectures and continues to publish regularly. She is a Fellow of the American Academy of Forensic Sciences and was a founding member of both NecroSearch International and the Ecology Section of the Botanical Society of America.

David O. Norris, PhD

Dr David Norris has done research in environmental endocrinology and neuroendocrinology for more than 50 years. Dr Norris is a professor emeritus in the Department of Integrative Physiology at the University of Colorado. He received his bachelor's degree from Baldwin Wallace College and his PhD in 1966 from the University of Washington. Dr Norris has worked in the area of forensic botany with Dr Jane H. Bock, since 1982, primarily on developing the use of plant cells in the gastrointestinal tract to aid in homicide investigations. Dr Norris and Dr Bock have been involved in investigations in numerous states as well as throughout the State of Colorado. Dr Norris has been certified as an expert witness in this area for the State of Colorado. With Dr Bock, Dr Norris also has consulted on other botanical evidence for criminal investigations. He was elected as a Fellow of the American Academy of Forensic Sciences in 2014 and also was a founding member of NecroSearch International.

Foreword by Tom A. Ranker

I was delighted when I received a request from my long-time friends and colleagues Jane Bock and Dave Norris to write a foreword to this book. For many years now I have heard various bits and pieces about the court cases that Jane and Dave contributed to as scientists, partly from a number of presentations they gave to my botany classes at the University of Colorado. Since people are always fascinated to hear how the application of scientific principles, observations, and analyses can be applied to criminal cases, I knew that I could always rely on them to give stimulating talks about plants and crime.

Jane Bock and Dave Norris have over 50 years of combined experience of applying sound scientific principles to help solve real crimes. They have blended their scientific specialties of plant ecology (Bock) and endocrinology (Norris) to form an impressive forensic scientific team that gathers, analyzes, and interprets a wide array of plant-based evidence from crime scenes, suspects, and victims. Thus, they are ideally situated to write a textbook on forensic plant science.

As a practicing plant taxonomist with experience applying data from plant anatomy and morphology, ecology, molecular systematics, and biogeography to basic scientific research, I appreciate the great attention to detail provided in this book and, in particular, on the emphasis of doing excellent science to provide the best possible evidence to help solve crimes. As a long-time herbarium curator, I also know the importance of "knowing your stuff" when called upon by local law enforcement to assist with the interpretation of botanical evidence. This book will not only help train novices in the field of forensic botany but also will assist experienced plant scientists and other professionals to apply botanical knowledge to criminal investigations.

Forensic Plant Science is particularly timely in light of the 2009 report of the National Academy of Sciences that decried the state of forensic science. One of the primary concerns expressed in that report was the lack of standard procedures employed across forensic labs, police departments, and jurisdictions. This book will help resolve this dilemma at least for plant forensic science by providing readers with (1) introductions to basic plant biology and the subdisciplines of botany needed for forensics, (2) actual examples of how plant-based evidence can and cannot be used in court, and (3) a critical "how to" manual for gathering, analyzing, and interpreting all sorts of botanical forensic evidence.

Tom A. Ranker, PhD, Professor
Department of Botany
University of Hawai'i at Mānoa
Past President, Botanical Society of America

Foreword by Haskell M. Pitluck

One of the perks you get when asked to write a foreword of a book is the galley copies of the unedited final draft to assist you to make an assessment which in this case is that this is a good book.

Forensic scientists have been working for years to assist our legal system in assuring that innocent people are not convicted and those who are guilty are convicted. Drs Jane Bock and David Norris have used over 30 years of experience in their field to author a book of eleven chapters with seven detailed appendices and citations. ***Forensic Plant Science*** is packed with excellent information to further the knowledge and use of plant science forensically in assisting the conclusion of legal cases, both civil and criminal. Their combined knowledge is an asset that they are sharing in a well-organized fashion with their readers.

The authors capture your attention from the first chapter with a basic introduction of plants as well as interesting cases with direction as to where to find evidence and how to present it in court.

The photographs and explanations are excellent. The appendices and online photomicrographs will be valuable tools to aid in the collection and processing of evidence. This book is a comprehensive study of not only plant science itself, but also of issues not directly related to plants. The information will assist in preparing for a legal matter involving plant science evidence.

Drs Bock and Norris discuss issues of plant science in the past, deal with present situations, and give insight into what may evolve in the future. Topics as diverse as the public's perception of forensic science and the "CSI effect" as well as how to get into the plant science field and the pros and cons of doing so.

In a relatively few short years, DNA has become the standard for positive identification. Plants have DNA as well, which will aid in the development of evidence. Studies of pollen and diatoms can be used to place people as well as items at a crime scene.

The authors also make a case for a forensic science professional society recognizing contributions by plant scientists to forensic science, including the certification of forensic plant scientists.

Whether or not that happens, the advances made in plant science will continue to bring a strong arrow in the quiver of those striving to find the truth in the legal system.

The authors are to be congratulated on producing a book that gives so much information in an uncomplicated way so as to be used and understood by investigators, attorneys, and judges.

Read it. Enjoy it. Learn from it.

Haskell M. Pitluck
Retired Circuit Court Judge, State of Illinois
Past President, American Academy of
Forensic Sciences 1995–1996

Preface

Jane Bock and David Norris first became friends while teaching General Biology together as young assistant professors in Boulder, CO. We enjoyed teaching together as we set about establishing our research careers. Bock was preoccupied with learning the Colorado flora and Norris with establishing a lab where he could work on the endocrinology of fishes and amphibians. Norris had a sound background in general botany and Bock knew about salamanders from fieldwork. Norris discovered that Bock knew very little about animal biology, so we formed a team teaching approach in which one did plant biology and the other covered the animals.

In General Biology, Norris described human digestion while Bock remained largely ignorant of human biology in general. A partnership was formed, Norris for human digestion and Bock for plant anatomy of food plants. Norris based his digestion lecture on human digestion of a Big Wally cheeseburger. Big Wally contents mimic those of a famous food chain's cheeseburger. Big Wally was born lest we run afoul of the big burger franchise by naming the lecture after their product.

One autumn day, Dr William (Ben) Galloway called Bock to ask if she could identify food plant cells from a murder victim's stomach contents. By this time Bock had moved on from General Biology to teaching Plant Anatomy and Plant Systematics while Norris was teaching Comparative Endocrinology and related subjects.

Because of the notoriety to Galloway's case and our contribution to its solution, regional police noticed us. We became involved in forming NecroSearch International, an organization that continues to lead in the search for clandestine graves. Bock and Norris spoke at some coroners' conventions, wisely joined the American Academy of Forensic Sciences and received a small grant from the Department of Justice. Soon they became associated with the general subject of botany by criminal investigators in the Front Range of Colorado. From the start of our collaboration we were asked to identify plant species in nature, food plant cells in the stomach of homicide victims, and share ecological knowledge concerning plant distributions. We continue to do this today, but the geographic distribution of our work has increased greatly. Our cases at first were from Colorado, but cases now come from other states and even outside the US.

Along the way we have developed working procedures including working independently when possible and then sharing our findings. We seek consensus whenever possible. Sometimes we are unable to answer questions or lack time to take on a new case because of other career demands. We try to give priority to child deaths and cases asking unusual questions such as "Can you tell from the last meal where (in what jurisdiction) the homicide took place?"

To spread information about our work, Norris and Bock have given lectures at colleges and scientific meetings throughout the US as well as in England, Australia, and New Zealand. We have given short courses at the conventions of the Botanical Society of America (BSA), the American Academy of Forensic Sciences (AAFS), and the Colorado State Police Academy as well as the Oregon State Police Forensic Laboratory. Especially rewarding was our recent course for high school science teachers from around the US that was sponsored by the AAFS and the University of Colorado, Boulder.

Sometimes when we receive requests for assistance, we solicit help from botanists of good reputation who are located nearer the crime scene. These cases involve questions of taxonomy and ecology and, in one case, identification of wood anatomy. We usually find people eager to help, although in a few cases people refused because they felt threatened by how our justice system deals with homicide investigations.

Of course, plant food cells can be identified past the stomach in the digestive tract and even outside the human body. Norris furthered our investigations by using crime scene fecal samples (on clothing) from a victim and a suspect in a rape homicide that inked that victim with a suspect. We also have identified food plants from vomitus samples. Bock serves on thesis defenses from Anthropology graduate students who study such subjects as plant cells associated with mummies and fecal remains from an old outhouse. Such studies can reveal dietary habits of people from past times. Numerous undergraduate students have worked in our labs on research projects related to forensic botany, and forensic botany has been a part of Norris's lecture and laboratory class, "Forensic Biology" at the University of Colorado. Bock integrated this subject into her botanical courses as well.

Bock's and Norris's investigative work comes from many sources. In the early work, computer literate police investigators sought help in some aspect of botany. In more recent times, word of mouth and alumni from Bock's and Norris's classes and our forensic publications have brought us inquiries.

An important goal Bock and Norris have for this book is to advertise to the legal community the value and efficacy of evidence from plant science. A second goal is to encourage those who have interest in or are trained in plant science to pursue forensic botany as a career.

Acknowledgments

In acknowledgments, the common practice is to mention family members last. That behavior does not fit here. We owe our mates, Carl E. Bock and Kay W. Norris, enormous gratitude for their patience and encouragement. We have lost count of the many dinner parties we ruined when preoccupied with a case or a forensic research question. The contents of this book show matter that is not a normally acceptable dinner conversation. And our daughters, Laura, Sara, and Linda suffered, too, not always with silence. These people endlessly make our lives worthwhile and rewarding.

We also wish to thank the people who inspired us to do this work:

Dr. William (Ben) Galloway, the forensic pathologist who started us on our life of crime
Jack Swanberg, the founder of NecroSearch International
Thomas Trujillo, Detective for the City of Boulder
Thomas Faure, former Boulder County Coroner
Tom (Grif) Griffin, Colorado Bureau of Investigation (retired)
Dorothy Sims, Esq.
Jose Baez, Esq.
Lawrence W. (Tripp) DeMuth, Esq.

The botanical colleagues we have worked with including:

Dr. Meredeth A. Lane
Dr. William (Ned) Friedman
Dr. Pamela Diggle
Dr. Yan B. Linhart

The students who helped us out in the laboratory, especially:

Scott G. Clarke
Janessa (Jacobs) Jacarrith
Collin Knaub
Laura Young
Mark Norman
Adelita Mendoza
Ryan Kuenning

The colleagues who contributed directly to this publication including:

Dr. Meredith A. Lane, Dr. Patrick Kociolek and Joshua Stepanek for providing scanning electron micrographs
Stephanie Mayer for access to the plant slide collection in the Department of Ecology and Evolutionary Biology, University of Colorado

Dr. Deane Bowers, Dr. Adrian Carper, and Virginia Scott for assistance in preparation of seed photomicrographs
Dr. Lee Reed of *NecroSearch International* and Dr. William (Ned) Friedman for use of photographs
Wendy Beth Jackelow for the artistic rendering of our crude drawings into colorful and useful illustrations

And especially Dr. Thomas Ranker and the Hon. Haskell Pitluck for writing the forewords.

And last, but not least, we wish to thank our Editor, Josyln Chaiprasert-Paguio and Senior Project Manager Lisa M. Jones with Elsevier Academic Press who made all of this happen.

CHAPTER

1

Introduction to Forensic Plant Science

Justice defined: "*The fair and proper administration of laws.*" *Black's Law Dictionary*, 9th edition. 2009.

The use of the word *forensic* related to crime is now especially popular because of contemporary media, in particular television. However, *forensic* has two definitions. The first is related to public speaking. School forensic clubs historically were debating societies. For our purposes, *forensic* applies to matters pertaining to courts and the law. Therefore, forensic plant science's definition is the application of plant evidence to legal questions. It is interesting that a number of aspects of forensic science are being debated in and out of the courtroom today.

Our purpose for this book is to show several aspects of plant science that have received little attention in the past but that can be especially useful in forensic science. Three of these areas are plant anatomy (Chapter 4), plant taxonomy (Chapter 6), and plant ecology (Chapter 8) that deal primarily with seed plants (e.g., flowering plants and conifers). Our forensic research, teaching, and casework are centered on these areas. Additionally, recent advances in genetic analyses of plants show promise for plant DNA-based forensics (Chapter 3). Lastly, the examination of diatoms (microscopic algae) and pollen (male reproductive sex cells) of seed plants as well as spores of some other plants are beginning to be developed as forensic tools (Chapter 10).

We have worked mostly on homicide cases, but plant science can be useful in the forensic analyses of rape cases, burglaries, and other crimes as we will describe in later chapters. For example, plant cells can help determine time of death through the analysis of gastrointestinal contents. Wood identification and comparisons can help identify a suspect. Plant fragments lodged in a shoe (Figure 1.1), associated with clothing, or found attached to or within a vehicle may link a suspect or a victim to a specific location. Vegetation analyses can be helpful in the location of bodies or clandestine graves. Diatoms may provide evidence of drowning and also can be used to characterize a location. Pollen of different species can help determine when or where a person was killed as well as connect suspects to crime scenes.

We illustrate in the following chapters for forensic scientists, crime investigators, and forensic science students how these different aspects of plant science are simple to use, can

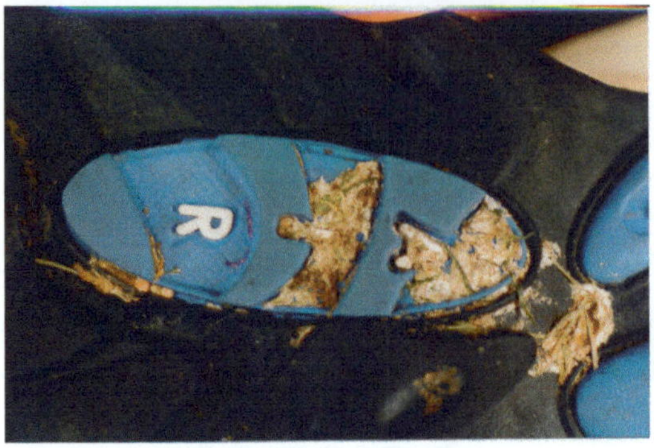

FIGURE 1.1 Plant material embedded in the tread on the bottom of a suspect's shoe. Identification of these plant fragments can connect a suspect to a specific site. *Photograph by author.*

be readily accepted in court, and, for the most part, are inexpensive. We hope also to interest practicing plant scientists and qualified students to pursue these avenues in forensics. In this introductory chapter, we must first provide an introduction to plant science and a little ancient history on the forensic aspects of plants.

1. INTRODUCTION TO PLANTS

Plant science is the branch of biology dealing with plant life. There remains little exact agreement about what organisms should be called *plants*. In this section, we deal primarily with seed plants, most of which carry out photosynthesis.

In the past few decades, this two-word term, *plant science*, has gained currency over its predecessor, *botany*. Plant science is the term used by the major governmental and private funding agencies for plant research today. Therefore, the use of botany to mean the same thing has declined. Perhaps, the word *botany* brings to mind "posey pickers," and some biologists and biochemists who work with plant materials possibly feel diminished when referred to as mere "botanists" rather than as "plant scientists."

The term "plant" usually refers to a great diversity of organisms varying from microscopic single cells to huge organisms such as the giant sequoia tree of California. Included are algae, bryophytes (e.g., mosses, liverworts), ferns, **conifers** and other **gymnosperms**, as well as the **flowering plants** or **angiosperms**. The term **seed plants** refers only to the gymnosperms and angiosperms that produce seeds. Our focus here will be on the flowering plants that dominate the terrestrial landscape, including other groups where they have achieved forensic significance.

1.1 The Seed Plant Body

Seed plants have only three organs, and you already know them. They are **leaves, stems**, and **roots** (Figure 1.2). These organs in turn are made up of tissues that are much simpler in comparison with those found in vertebrate animals. **Flowers** are the reproductive structures of angiosperms that are modified from leaves. Part of the flower will develop into a

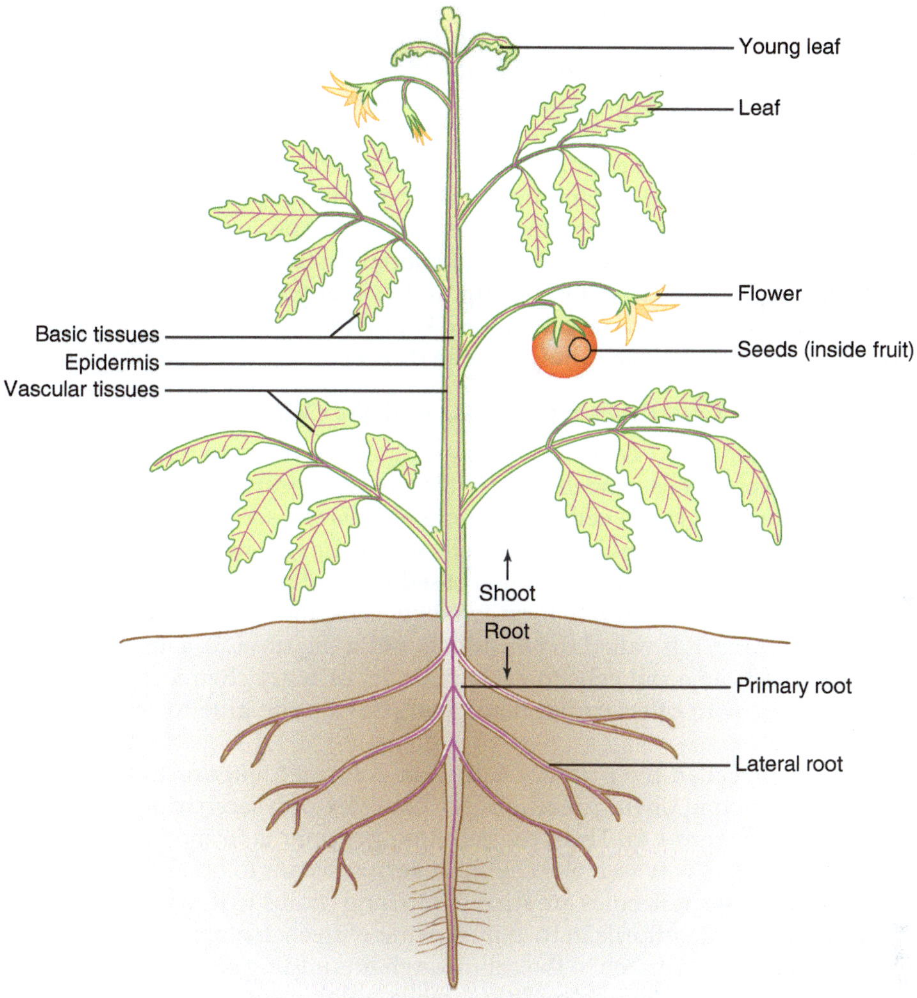

FIGURE 1.2 Organs of a flowering seed plant.

fruit that contains one to many seeds. Structurally, there are many parallels between roots and stems of both gymnosperms and angiosperms. The reproductive structures of conifers are called **cones**.

For the forensic work we describe here as **plant anatomy**, we deal only with plant cells. In forensic matters where **plant taxonomy** (identification) and **plant ecology** (plant interactions with their environment) are used, we deal primarily with entire plant organs along with certain other considerations such as **plant physiology** and **plant geography**. The study of plant form is called **morphology**. Plant morphology corresponds to what zoologists call "anatomy," whereas plant anatomy corresponds to cellular anatomy and histology in animals. Sound reviews of general botany are readily available for details about these areas (e.g., Mauseth, 2012; Raven et al., 2012).

1.2 The Seed Plant Cell

Our successful use of plant anatomy in homicide investigations depends on some knowledge of structural details of food plant cells (mostly angiosperms). Like animal cells, the plant's cell membrane surrounds the living contents of the cell, the **protoplast**. Within the protoplast are the cell's organelles, including the nucleus, mitochondria, Golgi apparatus, and ribosomes (as in animal cells) plus plastids including the chloroplasts and vacuoles found only in plant cells. **Vacuoles** are delimited within the protoplast by a special membrane, the **tonoplast** (Marty 1999); Figure 1.3. Vacuoles are multifunctional. In living cells, they are important in maintaining cell turgor as well as for storing and exchanging products of photosynthesis. Other metabolic byproducts such as crystals are kept within the vacuoles because they could interfere with normal metabolism if they were in direct contact with the protoplast.

1.2.1 A Unique Plant Constituent: Cellulose and the Cell Wall

Identification of plant food cells from human digestive tracts requires knowledge of the shapes and sizes of these cells as well as how they may appear together in plant fragments. Seed plant cells, as well as many cells of the nonseed plants, are enclosed by rigid cell walls external to the cell membrane, unlike animal cells where the cell membrane is directly exposed to the environment. The following account is based on seed plants.

As plant cells divide and mature to form two new cells, the first wall layer is formed external to the cell membrane. It is called the **middle lamella** (Figure 1.3). This layer forms during cell division and binds adjacent cells to each other. Its primary chemical content is pectin, a polysaccharide along with other components that provide the glue to cement plant cells to each other.

The next cell layer, called the **primary cell wall**, is formed interior to the middle lamella (Figure 1.3). Its major component is **cellulose**, a complex polysaccharide with the empirical formula $(C_6H_{10}O_5)n$ (Figure 1.4). Thus, the cellulose polymer is formed of glucose units that are connected in a unique way to make cellulose very resistant to breakdown.

Thousands of cellulose molecules are strung into long chains to form thin strands or microfibrils (Figure 1.5). The microfibrils in turn intertwine with each other making a sort of basket weave. Other molecules can attach to the cellulose chains adding strength to the cell wall. The primary cellulose cell wall, like the middle lamella, retains some flexibility, and its porous property allows for intercellular exchange of materials. At this time the new cell can increase in size.

Once the expansion in size of the primary wall is complete, almost no further change in size and shape of the cell takes place. Plant cells with completed primary cell walls may go on to specialize for specific functions. Cells that possess only the middle lamella and primary cell wall are called **parenchymal cells**. They make up much of a living plant's body. Parenchymal cells carry out many functions including photosynthesis and the transport and storage of the products of photosynthesis. They also are involved in movement of water and minerals from the soil into the plant body. Some parenchymal cells differentiate into other cell types such as **collenchyma** that has an extra thick wall that is flexible. For example, celery strings are composed of collenchyma (see Chapter 4 for details on basic plant cell types).

Once cell enlargement ceases, a **secondary cell wall** may be formed. Secondary walls are formed interior to the primary cell wall. Some secondary walls also are composed primarily of cellulose layers. Sometimes the secondary wall is made more rigid by the addition of **lignin**

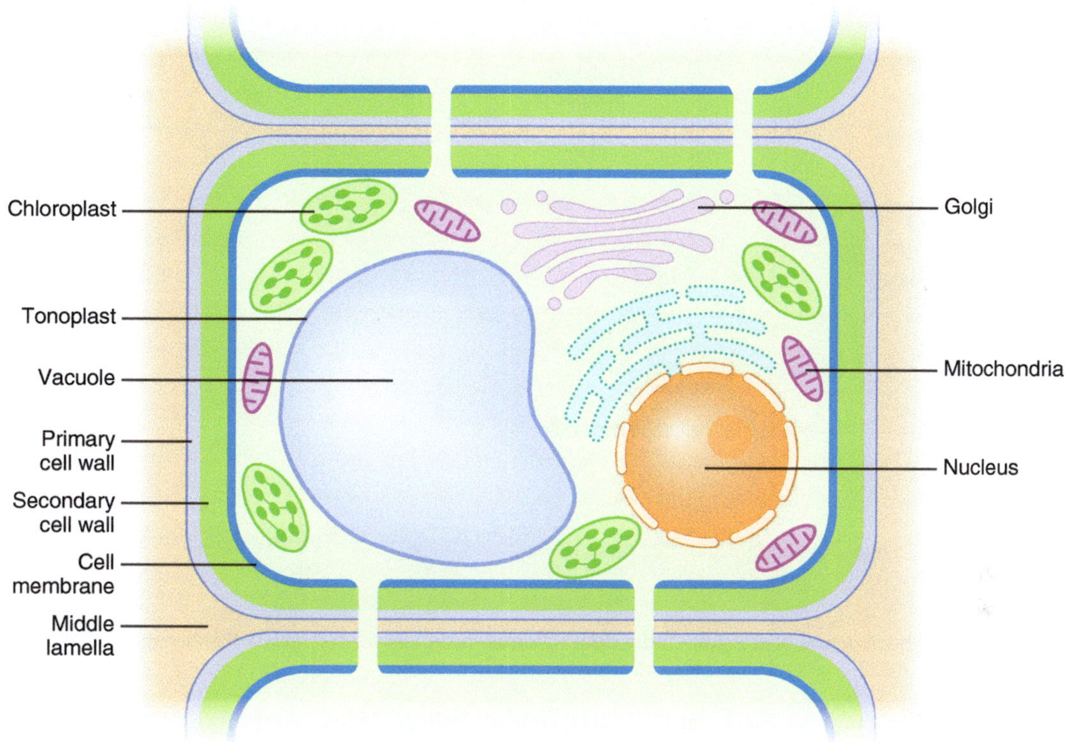

FIGURE 1.3 Generalized plant cell. Note the location of the middle lamella, primary cell wall, and secondary cell wall.

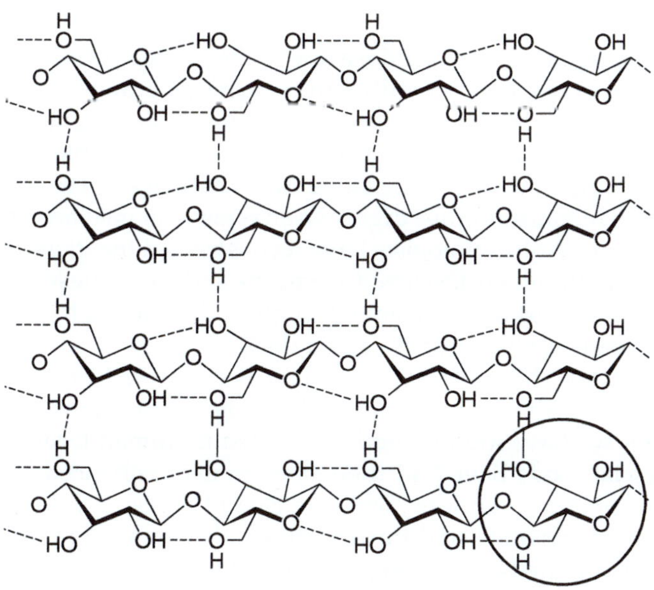

FIGURE 1.4 The generalized structure of cellulose showing hydrogen bonds (H–OH) connecting the glucose units (circle indicates one glucose molecule).

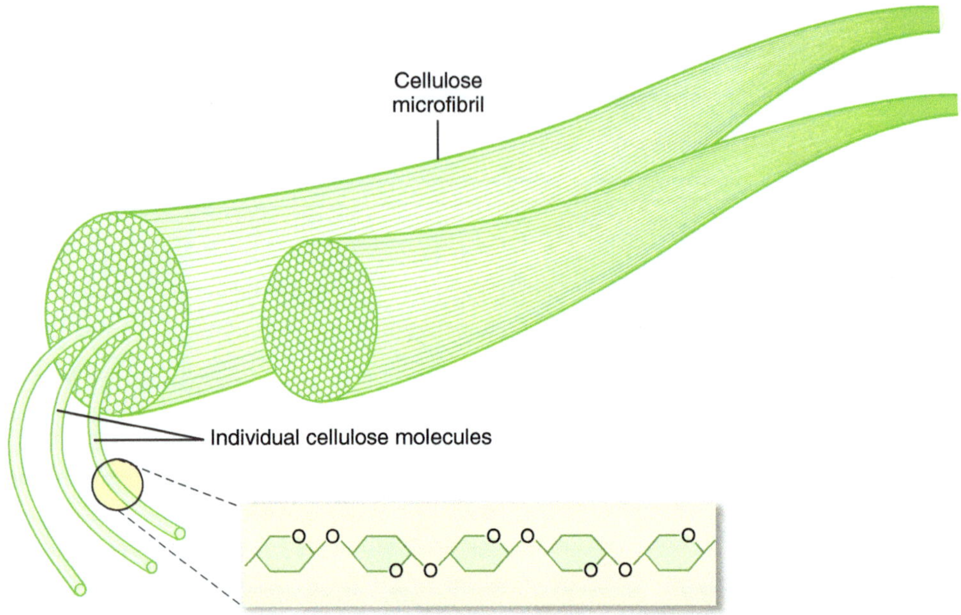

FIGURE 1.5 The structure of cellulose. Individual cellulose molecules form a matrix that becomes incorporated into a microfibril. Additional molecules may be added to the microfibrils.

molecules scattered among the cellulose microfibrils. These cells, usually dead at maturity, add strength and hardness to plants. Cells with lignified secondary walls are called **sclerenchyma**.

Cellulose maintains its shape and size under many conditions, including freezing, drying, boiling, and baking. This means plant cells maintain their distinctive shapes regardless of the methods of food preparation. However, the cellulose cell walls are destroyed by excessive grinding or burning.

Humans as well as most other animals cannot digest cellulose due to the unique pattern of glucose units and how they are connected to one another within the polymer. Certain microorganisms, however, can digest cellulose. These include microorganisms found in the guts of termites and the rumens of ruminate animals such as cattle. Because of the indigestible cell walls, many plant cell walls may pass through the human digestive tract unchanged in shape and size. Since the cell wall of food plants remains porous, digestive enzymes can gain access to the protoplast and digest the cell contents. Anthropologists have reconstructed the diets of ancient people by identifying plant cells within their fossilized stomachs and fecal intestinal contents. However, it was not until the latter part of the twentieth century that stomach contents and feces were used in criminal cases (see Chapter 5 for examples).

The presence of cellulose contributes to a "high-fiber diet." Wood is composed primarily of cellulose as are paper and cardboard. Although we normally do not consume wood and wood products for food, sometimes we have found wood embedded in human tissues. However, certain "high-fiber" commercial foods have had their fiber content enhanced by the addition of wood sawdust by unscrupulous vendors. Since we cannot digest cellulose, the sawdust adds no additional calories to our diets, and this practice is discouraged today. However, it is added to some prepackaged grated cheeses to prevent sticking.

2. THE EARLY HISTORY OF PLANT SCIENCE

The earliest uses of plants by humans were both agricultural and pharmacological. Early humans discovered by trial and error that some plants were edible but others were toxic. Still others proved to have beneficial effects as curatives and pain relievers. The earliest written records of forensic plant science deal with plants' curative and poisoning properties. Indeed, plant-derived poisons have a long history of criminal uses.

2.1 Pharmacology and Toxicology of Plants

Knowledge of plants that relieve suffering and cure illnesses along with plant poisons and their antidotes must have been learned through trial and error by early humans. That knowledge was passed on through intergenerational teaching. Today, over 75% of the world's people depend on herbal medicines (Simpson and Ogarzaly, 1995). But in the USA, only about 10% of the medical pharmacopeia comes directly from plants. However, many North American drugs are chemically synthesized from compounds found originally in plants (Simpson and Ogarzaly, 1995).

Records of plant uses abound in Sanskrit. Ancient Chinese medicinal records also are available. Shen Nung, the second Celestial Emperor (2000 BC), sampled more than 1000 herbs to determine their curative and poisonous properties. He died perhaps from sampling one poison too much (Magner, 1992). Simoons (1998) reviewed how poisoning showed up in diverse cultures, including the ancient Egyptians, the Greeks, and the Romans, continuing down to modern times.

The **Hippocratic Oath** originated around 400 BC. It remains a standard of values for many medical practitioners. Today, it is only administered in approximately 60% of U.S. medical schools at the granting of MD degrees (Jhala and Jhala, 2012). Older versions of the Oath forbade the use of poisons by physicians. This caution is not present in today's versions, although it may be covered by the caution to *do no harm*. Perhaps its omission relates to the treatment with many drugs that in higher concentrations could be lethal. Nevertheless, the use, even by medical practitioners, of plant-derived poisons in homicides became more common in the mid-nineteenth and early twentieth centuries as the chemical detection of metallic poisons such as arsenic improved (see Blum, 2011).

3. PLANT POISONINGS

During medieval times, accidental and intentional (i.e., homicides) poisonings with heavy metals (e.g., antimony, arsenic, others) were commonplace. Many of these poisons were active ingredients of curative potions and were readily available to people. Arsenic earned the nickname of "inheritance powder" as it was often used to hasten the postmortem transfer of property and wealth to heirs. However, by the nineteenth century, chemists were developing procedures for detecting these poisons making them less attractive for nefarious purposes. By the beginning of the twentieth century, there was a transition from the metallic poisons to alkaloids and other chemicals of plant origins that were more difficult to detect. Many of these plant poisons had been known for centuries and had made their way into many folk medicines.

Down to the present time, plant poisons and their derivatives continue to play an important role in forensic plant investigations. These can be the results of accidents or planning. In both modern medicine and quackery, plant products are used to treat serious medical problems, but such uses are two-sided because in excess these same plants can be fatal. One example of accidental plant poisoning came from drinking milk from Indiana cows that had fed on white snakeroot (*Eupatorium rugosum*, family: Asteraceae). The disease known as "milk sickness" killed Nancy Hanks Lincoln, Abraham Lincoln's mother. She was responsible for his learning to read and write. Lincoln said (Holland, 1866), "All that I am, or hope to be, I owe to my angel mother—blessings on her memory." Carlier et al. (2014) points out that plant poisonings remain common.

3.1 Some Specific Poisons of Plant Origins

We have selected some of the more notorious plant poisons to describe: alkaloids, glycosides, and lectins. All of these toxins were discovered as having some sort of curative property by ancient people, and many of them are still used today. General symptoms of poisonings caused by these compounds are summarized in Levine et al. (2011).

3.1.1 Alkaloids

Alkaloids are chemicals that contain basic nitrogen atoms. They consist mostly of carbon, hydrogen, and nitrogen but may also contain sulfur and/or oxygen. Rarely, they will include elements such as chlorine, bromine, or phosphorus.

3.1.1.1 COLCHICINE

One plant poison favored by the Greeks and Romans came from a species of crocus (*Colchicum* spp., family: Iridaceae). These plants are the source of the alkaloid drug **colchicine** (Figure 1.6) that sometimes is prescribed today for the treatment of gout, arthritis, and constipation-predominant irritable bowel syndrome. Colchicine is known to most biologists as an inhibitor of cell division. The drug comes from crocus corms and seeds. Colchicine can be deadly if misused as there is no known antidote for colchicine poisoning. Multiple system failures occur in 24–72 h after consumption of a lethal dose. An unfortunate case of colchicine poisoning occurred in Colorado when a thief misread the label on a bottle he had stolen from a doctor's safe, and he died after ingesting the pills (Bock, personal observation).

3.1.1.2 POISON HEMLOCK

Another plant whose poisonous properties have been well known since ancient Greece, is **poison hemlock** (*Conium maculatum* L.), a member of the carrot family (Apiaceae). Important people in ancient Greece who received death sentences were allowed to choose their method of death. Socrates selected poisoning with a tea made from poison hemlock. His devoted student Plato witnessed his death and described in detail the stages of the poison's action (Gallop, 2009). Plato's description fits the symptoms of contemporary poisoning by poison hemlock (Lewis and Elwin-Lewis, 2003). The active toxic ingredient is the alkaloid **coniine** (Figure 1.6), which causes paralysis of the respiratory muscles leading to death. As little as 100 mg (1.6 mg/kg body weight for a 60 kg adult) is a lethal dose (e.g., six to eight leaves of

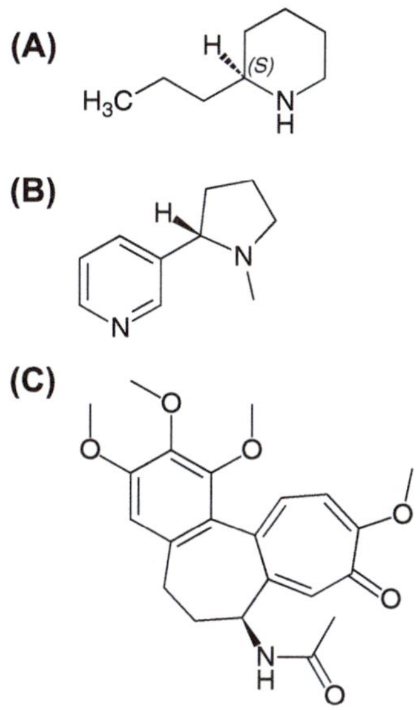

FIGURE 1.6 Alakaloid toxins. (A) Coniine; (B) Nicotine; (C) Colchicine.

C. maculatum). Poison hemlock is widespread in the northern hemisphere and a few cases of poisoning occur each year.

3.1.1.3 THE TROPANE ALKALOIDS

Deadly nightshade, *Atropa belladonna*; jimson weed or "loco weed" (*Datura* spp.); angel's trumpet (*Brugmansia* spp.); and henbane (*Hyoscyanus niger*) are the sources of several hallucinogenic and potentially lethal tropane alkaloids scopolamine, atropine, and hyoscyamine (Figure 1.7). All of these plants are in the family Solanaceae that includes potatoes and tomatoes.

Scopolamine is used as an antidepressant and antinausea drug. It is anticholinergic and antimuscarinic. Paradoxically, overdoses can produce depression. It is hallucinogenic but the experiences are generally extremely unpleasant. Scopolamine at one time was administered to pregnant women in labor as "twilight sleep."

Atropine is also an anticholinergic, antimuscarinic drug that causes pupil dilation, increases heart rate, and increases secretion of saliva. A fatal dose of atropine is greater than 10 mg, whereas scopolamine is toxic at 2–4 mg. The name "belladonna" comes from Italy where it was once used to dilate the eyes of women to make them more attractive ("bella") to men.

Hyoscyamine is the levorotatory isomer of atropine and is also the precursor for the synthesis of scopolamine. Its actions are similar to scopolamine and atropine. Hyoscyamine is named for the genus of henbane that concentrates tropane alkaloids in the leaves and seeds (Figure 1.8).

Perhaps a leader in poisonings among these plants is jimson weed. This plant is featured in the many controversial books by Carlos Castenada (http://en.wikipedia.org/wiki/Carlos_Castaneda) that first came to prominence in the 1960s. Jimson weeds are very halucinogenic and can be fatal.

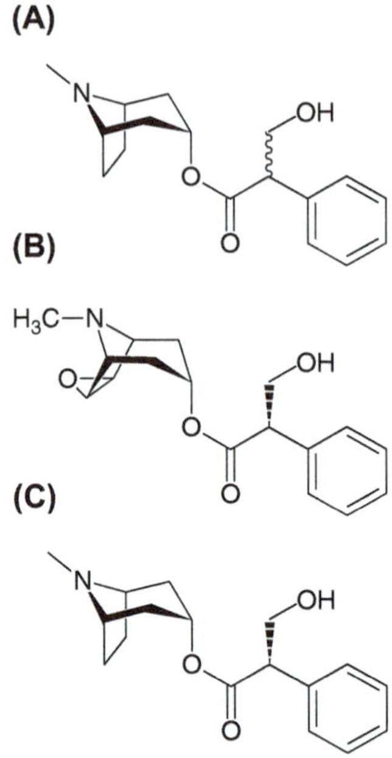

FIGURE 1.7 Structures of tropane alkaloids. (A) Atropine; (B) Scopolamine; (C) Hyoscyamine.

FIGURE 1.8 (A) Deadly nightshade, *Atropa belladonna*. *(Courtesy of Kurt Stüber, available at* caliban.mpiz-koeln.mpg.de/mavica/index.html *part of* www.biolib.de.*)* (B) Seeds of henbane, *Hyoscyanus niger*. *(Courtesy of Steve Hurst, USDA.)*

3.1.2 Other Alkaloids

3.1.2.1 STRYCHNINE

Strychnine (Figure 1.9(A)) is a potent alkaloid neurotoxin that blocks cholinergic receptors in skeletal muscles. Excessive doses can lead to paralysis of respiratory muscles causing asphyxia and death. The lethal dose for humans is 32 mg/kg body weight. Strychnine is

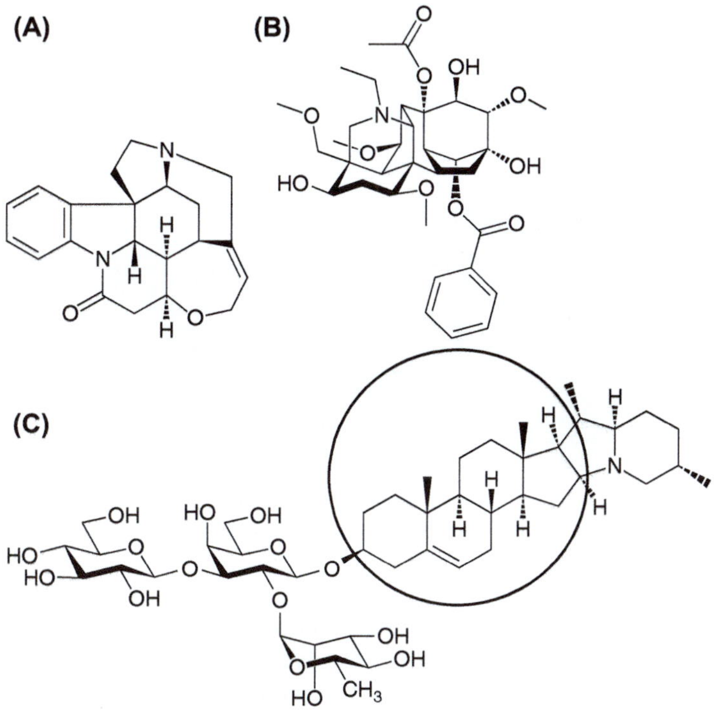

FIGURE 1.9 Some other alkaloid toxins. (A) Strychnine; (B) Aconitine; (C) Solanine. Note the inclusion of a steroid nucleus (circle).

most commonly derived from the seeds of *Strychnos nux-vomica*, a native tree of South India, and a related climbing shrub, *Strychnos ignatii*, native to the Philippines called St. Ignatius' bean. It was used as a rodent poison in Europe beginning in the seventeenth century until the present time. Accidental poisonings were not uncommon. It is suggested that the death of Jane Stanford in 1905, a cofounder of Leland Stanford University was a result of strychnine poisoning although it is not clear how this came about (Cutler, 2003). Once strychnine became readily available, it made its way into use for homicides. It has been suggested that it was the poison given to Alexander the Great in 323 BC (Phillips, 2004). However, because of the overt symptoms of strychnine poisoning and its easy chemical detection, it is not the poison of choice today. Nevertheless, it did appear in the San Diego death of Sue Morency who died under unusual circumstances in 1990. She had a body concentration of strychnine that was 4 times the lethal level. Her husband was arrested and charged with the homicide (Bellandi, 1990).

3.1.2.2 ACONITINE

Aconitine (Figure 1.9(B)) is produced by the 250 species of *Aconitum* commonly called **monkshood** (Figure 1.10). All parts of these plants are extremely toxic, especially the roots.

In addition to its use in ancient medicines, it was used to make poisoned arrows for hunting (Chinese, Japanese Ainu, Aleuts) and warfare (Chinese).

FIGURE 1.10 Monkshood, *Aconitum variegatum. Aconitum variegatum, Härtsfeld, Germany, courtesy of Bernd Haynold, available at* http://commons.wikimedia.org/wiki/File:Aconitum_variegatum_110807f.jpg.

Aconitine effectively opens sodium channels so that muscles and neurons cannot be repolarized. Thus, aconitine can produce ventricular dysrhythmia of the heart leading to death. It can also cross the blood–brain barrier and produce neural effects. One of the early uses of *Aconitum* extracts in Europe was to kill wolves, hence another of the plants' many common names is wolf's bane. The lethal dose for humans is 32 mg/kg body weight. Surprisingly, the caterpillars of numerous moth species feed on this plant despite its toxicity to many other animals. The lowest oral dose reported to kill a human is only 29 μg/kg body weight (100× more lethal than strychnine).

Reportedly, Cleopatra used aconitine to poison her brother (and husband) Ptolemy XIV so she could replace him with her son (http://en.wikipedia.org/wiki/Aconitum). A promising young Canadian TV and film actor, Andre Noble, died after consuming monkshood while on a hike in Newfoundland (Gallagher, 2004). In 2009, the British "Curry Killer," Lakhvir Singh, murdered her lover by feeding him a curry dish laced with aconitine (BBC, 2010).

3.1.2.3 SOLANINE, A GLYCOALKALOID

Potatoes (*Solanum tuberosum*, Solanaceae) that show signs of greening, sprouting, rotting, or physical damage should not be eaten because of the high concentrations of **solanine** (Figure 1.9(C)). If one observes green material beneath the skin of a potato, one should not eat the potato because solanine is concentrated in this green layer and there may also be elevated levels in the rest of the potato. Greening in a potato is evidence of excessive exposure to light. Solanine, like other cyanide compounds, is produced as a deterrent to insects and other animals that might feed on the plants. It is found at lower amounts in other food plants such as eggplant and green peppers.

In the USA, each adult human consumes about 65 kg of potatoes/year. Ingestion of potatoes high in solanine and a closely related glycoalkaloid, **chaconine**, has been associated with

TABLE 1.1 Toxic Effects of Potatoes (*Solanum tuberosum*) Containing Large Amounts of Solanine/Chaconine in Humans[a]

Affected	Potato type	Quantity consumed	Concentration of TGA[b] (mg/kg bw)	Estimated toxic dose (mg/kg bw)	Outcome
56 (soldiers)	Peeled, cooked (whole uncooked)	1–1.5 kg	24 (38)	3.4–5.1	Recovered
60 adults 1 child	Potatoes	500 g? 200 g?	41	3.4 4.5 (lethal)	Recovered 1 fatal (5 year old)
7 (family)	Greened potatoes	?	?	?	2 fatal
50–60 (Cyprus)	Shoots, leaves	?	27, 49	?	1 fatal
Prisoners	Experimental	?	?	2.8	Recovered
Child	Potato berries	?	?	?	1 fatal
4 (family adults)	Baked potatoes with skin	1–3 potatoes 150–450 g	50	1.2–3.2	Dose-related recovered
78 (schoolboys)	Old potatoes	2 small potatoes 200 g	25–30	1.4–1.6	3 comatose, all recovered; young boys, more affected
61 (school children) Alberta	Baked potato	200 g	49	2.5	Recovered

? = not determined.

[a] *Modified from Kuiper-Goodman, T. and Nawrot, P.S. Solanine and Chaconine. http://www.inchem.org/documents/jecfa/jecmono/v30je19.htm.*

[b] *TGA, toxic glycoalkaloids.*

numerous poisonings and some fatalities (Morris and Lee, 1984; see also Table 1.1). The compounds can cause neurological impairments, vomiting, and diarrhea. Most varieties of potatoes contain less than 5 mg/kg. Concentrations of 14 mg/kg potato cause a bitter taste and 20 mg/kg causes a burning sensation in the mouth and throat.

3.1.3 *Glycosides*

Glycosides are formed between a sugar (saccharide) and another functional chemical group. The glycosidic bond joining these components is usually formed through an oxygen, sulfur, or nitrogen atom. A glycoside with a sulfur bond would be a thioglycoside, for example.

3.1.3.1 DIGOXIN, A CARDIAC GLYCOSIDE

The family Solanaceae does not have a corner on poisonous/medicinal plants. There are about 20 species of **foxglove** (*Digitalis* spp.: figwort family, Scrophulariaceae). **Digoxin** (Figure 1.11) is a cardiac glycoside extracted from foxglove. It often goes under the name of **digitalis**. Some cardiac patients under treatment for congestive heart failure and atrial arrhythmia carry a supply of digitalis pills for self-medication if they feel symptomatic and

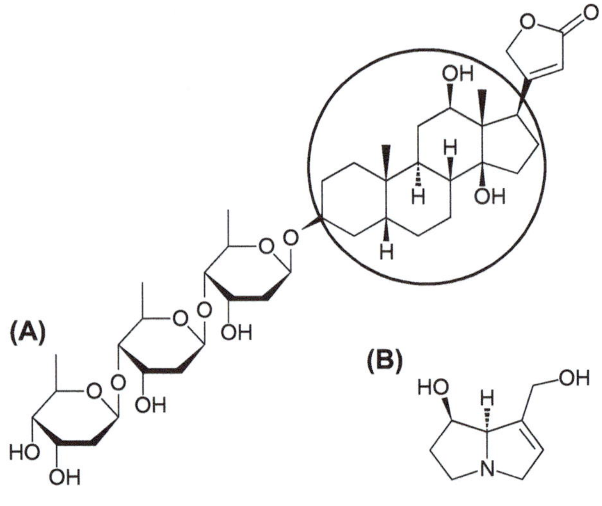

FIGURE 1.11 (A) Digoxin (digitalis), a cardiac glycoside from *Digitalis* spp. Note the inclusion of a steroid nucleus (circle). (B) Retronecine, a pyrrolizidine alkaloid extractable from comfey.

FIGURE 1.12 (A) Foxglove (*Digitalis* spp.). *(Digitalis purpurea, courtesy of Kurt Stüber, available at* caliban.mpiz-koeln.mpg.de/mavica/index.html *part of* www.biolib.de.*)* (B) Comfrey (*Symphytum* spp.). *(Symphytum × uplandicum, courtesy of en.Sannse, available at* http://upload.wikimedia.org/wikipedia/commons/0/0b/Russian_comfrey_close_800.jpg.*)*

are away from professional help, although its use is declining. Overdosing of digitalis can result in death. People have sometimes confused foxglove with **comfrey** (*Symphytum* spp.) and brewed a toxic "comfrey tea" (Figure 1.12). However, comfrey contains the pyrrolizidine alkaloid **retronecine** (Figure 1.12) that is hepatotoxic and linked to liver cancer and probably should not be ingested. Treatments for accidental and purposeful overdoses of these drugs remain a major field of research (Levine et al., 2011).

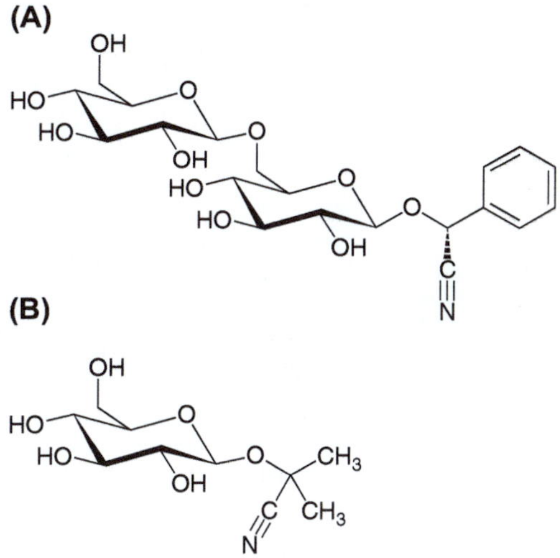

FIGURE 1.13 Cyanogenic glycosides. (A) Amygdalin; (B) Linamarin. CN = cyanide grouping.

3.1.3.2 CYANOGENIC GLYCOSIDES

Cyanide has been a popular poison for homicides and mass murders. It inhibits the mitochondrial enzyme cytochrome c oxidase and stops cellular respiration leading to death in minutes. It can be administered as a gas (**hydrogen cyanide** or **prussic acid**) or taken orally in compound form (e.g., potassium cyanide, sodium thiocyanate). In 2013, Urooj Khan of Chicago cashed in his lottery ticket for $600,000 only to fall ill the next day and die. It was considered a natural death until a relative pressed the authorities to exhume his body and do a toxicology scan. The results indicated that he had been poisoned with cyanide.

Although the common sources for cyanide are artificial, more than 1500 species, mostly angiosperms, produce **cyanogenic glycosides** as predator deterrents. Although these compounds may produce unpleasant effects in humans, the concentrations are not likely to be lethal. During droughts, the lack of water increases the concentrations of cyanogenic glycosides, and they prove more toxic to insects and other predators that attempt to feed on the leaves, stems, or roots.

Amygdalin (Figure 1.13(A)) is present in the almond fruits of *Prunus dulcis*. Its name is derived from the ancient Greek word for "almond." There are two varieties of almonds, one that tastes sweet (variety *dulcis*) and one that is bitter (variety *amara*, also called bitter almond). In the bitter almond, amygdalin is enzymatically converted to the toxic prussic acid and benzaldehyde, the chemical that gives the almond a bitter taste. The edible almonds consumed in the USA are sweet almonds, but bitter almonds can be found in specialty stores.

Cyanogenic glycosides may be found throughout many edible fruits, including, apples, peaches, pears, raspberries, cherries, apricots, and plums, but they are especially concentrated in the seeds. If these seeds are swallowed, they generally pass through the digestive system untouched. It is strongly recommended, however, that elderberries not be eaten raw as cooking releases a considerable amount of cyanide from the pulp of the fruits that diffuses harmlessly into the air. Bamboo shoots have high concentrations of cyanogenic glycosides as

FIGURE 1.14 Sources of lectin toxins. (A) Seeds of castor beans, *Ricinus annus*, source of ricin. (B) Seeds of jequirity, *Abrus precatorius*, source of abrin. *Castor beans, courtesy of HediBougghanmi2014, available at* http://en.wikipedia.org/wiki/Ricin#/media/File:Castor_beans1.jpg. *Abrus precatorius* Nutt. Ex Hook, courtesy of Steve Hurst, USDA.

do cassava roots. The processing of cassava roots (the source of tapioca) requires release of cyanide from the glycoside **linamarin** (Figure 1.13(B)) (also present in lower amounts in lima beans and flax). Otherwise, the cassava roots are very poisonous. Considerable amounts of cyanide are also released from burning tobacco although the amounts of cyanide inhaled are well below the lethal range.

3.1.4 *Toxic Plant Lectins*

Two very toxic plant **lectins** (carbohydrate-rich proteins or toxalbumins) are ricin and abrin. **Ricin** is made in the endosperm tissue of castor beans (*Ricinium* spp.). It is very toxic if inhaled or ingested (lethal dose = 22 μg/kg body weight) but considerably less toxic orally (lethal dose 1 mg/kg). Ricin inhibits protein synthesis but is often not fatal if treated. Castor beans (Figure 1.14(A)) are compressed into "castor cakes" that are high in protein (43%) and are used for organic fertilizer. These cakes are not appropriate for animal feed due to their high ricin content. Ricin has been used in assassinations, has been a candidate for a chemical weapon, and has been used to contaminate letters sent to political figures in the USA.

A major source of **abrin** is the jequirity, *Abrus precatorius*, an invasive pan-tropical plant originating in India. Abrin also inhibits protein synthesis but is much more toxic than ricin. The median adult human toxic dose orally is 10–1000 μg/kg, whereas the inhalation toxic dose is 3.3 μg/kg. Seeds (Figure 1.14(B)) of *A. precatorius* are often used as beads in jewelry.

3.1.5 *Dicoumarol and Anticoagulants*

Strychnine was replaced as a rat poison by the very effective warfarin in 1948. In the 1920s, some cattle developed a disease that caused them to bleed to death. It was discovered that the disease resulted from feeding cattle spoiled silage made from sweet clover hay. Sweet clover produces a nontoxic sweet-smelling compound called **coumarin** that certain fungi in the silage can metabolize into **dicoumarol** (Figure 1.15(B)), a potent anticoagulant that was responsible for the cattle bleeding to death. In the presence of dicoumarol, their blood would not clot. Researchers at the University of Wisconsin modified dicoumarol to produce **warfarin** (Figure 1.15(C)), which was an even more potent anticoagulant. Because after years of use, many rat populations became resistant to warfarin, chemists developed a highly lethal anticoagulant

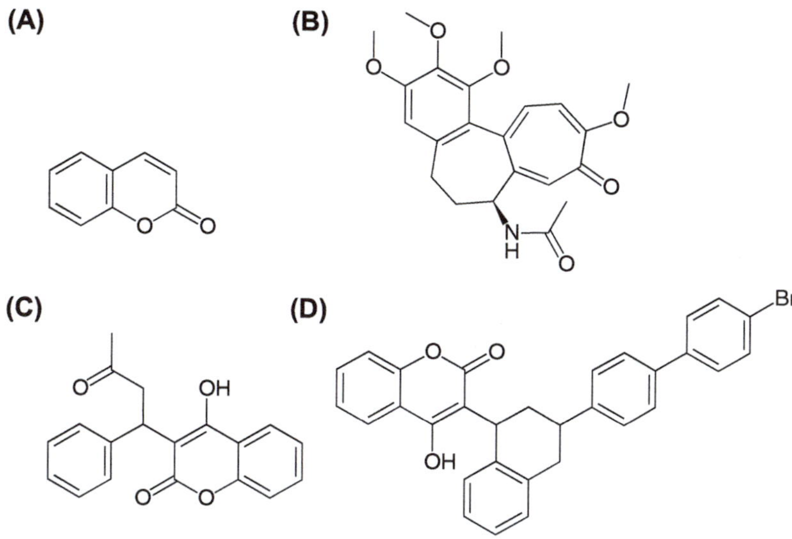

FIGURE 1.15 Anticoagulants derived from sweet clover. (A) Coumarin, the parent compound; (B) Dicoumarol, synthesized from coumarin by fungi; (C) Warfarin, a synthetic anticoagulant made from dicumarol in the laboratory; (D) Brodifacoum or "superwarfarin."

called **brodifacoum**, which is currently being used to poison rats as well as other pest mammals such as possums. The new drug, brodifacoum (Figure 1.15(D)), is sometimes called "superwarfarin." The 96-h LC_{50} for brodifacoum in rainbow trout (lethal concentration to kill 50% after exposure for 96 h) is a concentration of only 40 µg/L. The lethal dose for a 60-kg human male is 15 mg (250 µg/kg). Warfarin is sometimes administered today to humans as an anticoagulant.

3.1.6 Mushroom Toxins

Most cases of mushroom poisonings are caused by people collecting wild mushrooms and mistaking poisonous mushrooms for edible species (Levine, 2011). Occasionally, they are involved in homicides. Best known are the **amatoxins** found in several genera including *Amanita*. Amatoxins (Figure 1.16(A)) disrupt protein synthesis by inhibiting the enzyme RNA polymerase II. When ingested, the liver is the organ usually affected first and survivors may require a liver transplant. The estimated lethal dose for an adult human is about 100 µg/kg body weight.

Muscimol (Figure 1.16(B)) is a psychotic alkaloid found in some species of *Amanita* but is much less toxic than the amatoxins. It is a potent agonist of GABA receptors and causes visual perception problems and auditory hallucination. The LD_{50} (lethal dose to kill 50% of the test animals) for muscinol in mice is 3.8 mg/kg body weight.

Oreleanine (Figure 1.16(C)) is a nephrotoxic bipyrridine dioxide isolated from *Cortinarius* spp. The LD_{50} for mice is quite high (12–20 mg/kg body weight) but it is believed that humans are more sensitive to oreleanine than mice. There is no known antidote.

Methylhydrazine (MMH) is the toxic substance in false morels (*Gyromitra* spp.). NASA used MMH as a rocket propellant in the Apollo lunar modules. Although it causes gastrointestinal upsets and is a potential carcinogen, it is usually not fatal.

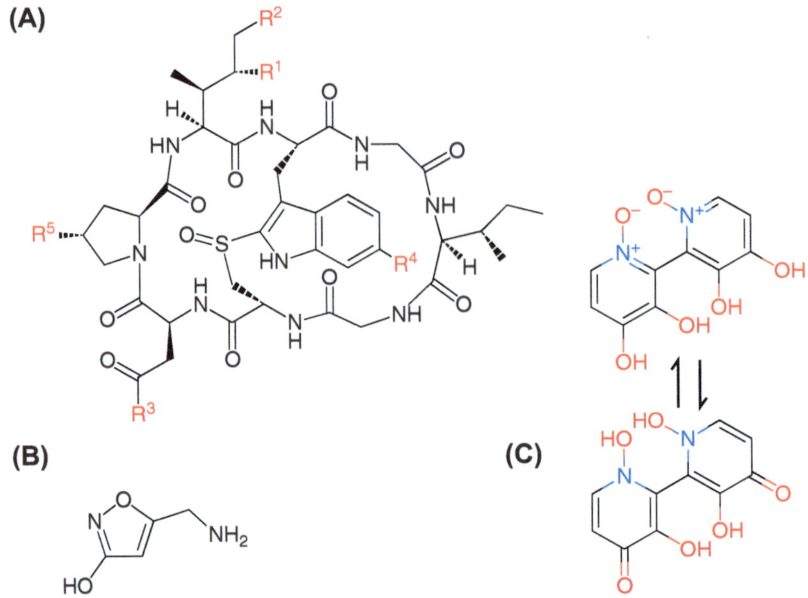

FIGURE 1.16 Some mushroom toxins. (A) Amatoxins, 10 are known with different substitutions at the "R" positions; (B) Muscimol; (C) Oreleanine.

4. ILLEGAL DRUGS OF PLANT ORIGINS

The illegal drug trade is centered around drugs that are addictive from plants that can readily be cultivated, extracted, and purified with minimal expenditures. The drugs are capable of producing euphoria and hallucinations and are lethal in high dosages. The forensic plant scientist may be asked to identify plants grown for this illegal trade by law enforcement.

The major drug trade has traditionally been focused on the products of the **opium poppy**, *Papaver somniferum* (Figure 1.17(A)). Knowledge of the pain-relieving capacity of this plant apparently extends from the Stone Age. The latex collected from the opium poppy contains three main addictive pain relievers: **morphine, codeine, thebaine** (Figure 1.18). Morphine and codeine are used medically as analgesics. Additionally, morphine is treated chemically to produce **heroin**, which has twice the potency of morphine.

Cannabis spp. are sources of **marijuana**. The principal psychoactive constituent in *Cannabis* is **tetrahydrocannabinol** (THC) (Figure 1.19(A)). Marijuana is used both recreationally for its psychoactive properties and medically for its analgesic properties. In the USA, at the time of this writing, marijuana is legal for medical purposes in more than half of the states and is sold for recreational use in two states (Colorado and Washington). Because THC can impair automobile drivers similar to alcohol use and is not considered to be legal in most states except for medical usage, marijuana legalization is a headache for law enforcement.

The use of psychoactive alkaloids obtainable from cacti is generally illegal as well. However, the use of the cactus known as **peyote** (*Lophophora williamsii*) (Figure 1.17(B)) that contains the alkaloid **mescaline** (Figure 1.19(B)) has historically been used ritualistically and

FIGURE 1.17 (A) Fruits of the opium poppy, *Papaver somniferum*; (B) Peyote, *Lophophora williamsii*, source of mescaline. *Photographs by author.*

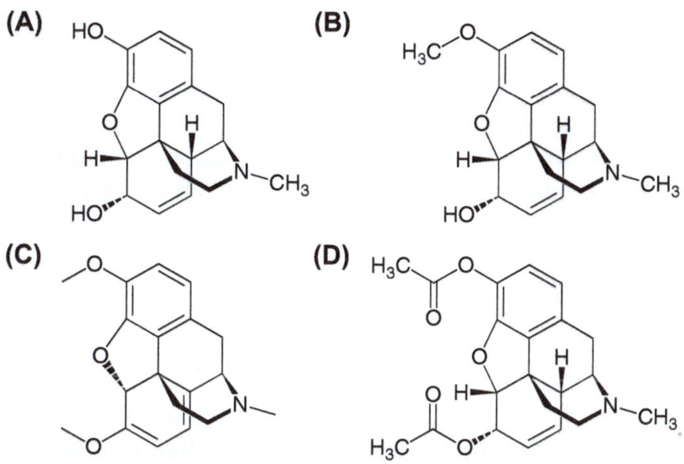

FIGURE 1.18 Natural opiates and heroin. (A) Morphine; (B) Codeine; (C) Thebaine; (D) Heroin.

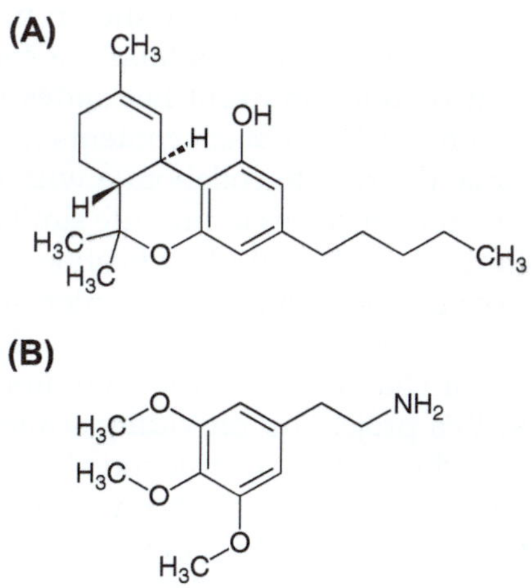

FIGURE 1.19 (A) Tetrahydrocannabinol from *Cannabis* spp.; (B) Mescaline from Peyote, *Lophophora williamsii*.

medically by indigenous people of southwestern USA and is considered an exception. Other psychoactive compounds are found in many cactus species, some of which are very toxic as unfortunately discovered by people who have experimented on themselves.

5. TWENTIETH-CENTURY FORENSIC PLANT SCIENCE

The first utilization of modern forensic plant science was in the area of plant taxonomy. The specific identification of poisonous plants as well as plants that were sources of illegal drugs became important when attempting to obtain convictions for possession and/or cultivation of the plants. However, it was the 1932 kidnapping and murder of the 20-month-old son of Charles and Anne Lindbergh that brought forensic plant science to the attention of the American public (PBS, 2013; Miller, 1994). Charles was a famous aviator and a national hero for being the first person to fly solo from the US to France. The Lindbergh child was abducted from a second floor nursery by the use of a crudely constructed wooden ladder that was left at the scene. Two years later, Bruno Hauptmann was arrested for the kidnapping after a portion of the ransom money was discovered in his possession. Hauptman claimed that the money was left with him by a former associate and that he had no idea it was connected to the kidnapping. However, a wood expert, Arthur Koehler, matched the grain in wood samples from Hauptmann's attic to the wood of the ladder used in the abduction. Kohler's analysis confirmed the wood from the ladder was from Hauptmann's attic and that toolmarks found on the wood pieces matched marks left on test wood by Hauptmann's tools. Replicas of the crude ladder were sold to attendees of Hauptmann's trial. He was convicted and sentenced to death.

6. OUR INTRODUCTION TO FORENSIC PLANT SCIENCE

Our first experience with forensic science came in 1982 when one of us (Bock) was contacted by Ben Galloway (Dr William B. Galloway) who at that time was a medical examiner for Jefferson County in Colorado and a professor of Pathology at the University of Colorado Health Sciences Center in Denver. Ben had a collection of materials from the stomach of a homicide victim that did not match the victim's last known meal, but he was unsure as to how to identify it. Galloway ascertained that Bock taught a course titled "Plant Anatomy" at the University of Colorado at Boulder and sent her slides to examine in order to positively identify the plant material from the stomach contents (for case details, see Chapter 5, pp 85-86). Bock asked a colleague (Norris) to collaborate with her in this work. Soon, Bock and Norris were being asked by other agencies to provide similar information. This led them to develop procedures for the examination and identification of plant cells and tissues from common food plants. They wrote a manual including a microscopic atlas of numerous food plants that was published by the National Institutes of Justice (NIJ) in 1988 (Bock et al., 1988). The NIJ distributed copies free of charge to forensic laboratories throughout the USA. Dr Meredith Lane participated in this project by providing scanning electron micrographs of some of the plant foods and helped with the construction of a key for identifying food plants from their microscopic structure. That manual has been out of print for more than 20 years, which was one of the motivations for writing this book.

Further Reading

General Topics

General Botany

Evert, R., Eichhorn, S.E., 2012. Raven Biology of Plants, eighth ed. W.H. Freeman.
Mauseth, J., 2012. Botany: Introduction to Plant Biology, fifth ed. Jones and Bartlett Learning.
Raven, P.H., Evert, R., Eichhorn, S.E., 2012. Biology of Plants, eighth ed. W.H. Freeman.

Plant Anatomy

Evert, R.F., Eichhorn, S.E., 2006. Esau's Plant Anatomy: Meristems, Cells, and Tissues of the Plant Body: Their Structure, Function and Development. John Wiley & Sons, New York.
Winton, A.L., Winton, K.B., 1935. The Structure and Composition of Foods. Vol. 2., Vegetables, Legumes, Fruits. John Wiley & Sons.

Plant Identification

Cullen, J., 2006. Practical Plant Identification: Including a Key to Native and Cultivated Flowering Plants in North Temperate Regions. Cambridge University Press, Cambridge GB.

Ecology

Cain, M.L., Bowman, W.D., Hacker, S.D., 2011. Ecology, second ed. Sinauer, Sunderland, MA.
Gurevitch, J., Scheiner, S.M., Fox, G.A., 2006. The Ecology of Plants, second ed. Sinauer, Sunderland, MA.
Smith, T.M., Smith, R.L., 2012. Elements of Ecology, eighth ed. Benjamin Cummings, New York.

Plant Toxins

Levine, M., Ruha, A., Graeme, K., Brooks, D., Canning, J., Curry, S., 2011. Toxicity in the ICU part 3: natural toxins. Chest 140, 1357–1370.
Roberts, F., Wink, M., 1998. Alkaloids: Biochemistry, Ecology, and Medicinal Applications. Springer.

References

BBC, February 10, 2010. Poisoning in West London in 2009. BBC TV News.
Bellandi, D., August 25, 1990. Husband Arrested in Woman's Poisoning Death. Los Angeles Times. http://articles.latimes.com/1990-08-25/local/me-818_1_strychnine-poisoning.
Blum, D., 2011. The Poisoner's Handbook. Penguin Books, New York.
Bock, J.H., Lane, M., Norris, D.O., 1988. The Use of Plant Cells in Forensic Investigation. U.S. Department of Justice, National Institutes of Justice, pp. 130.
Carlier, C., Guitton, J., Romeuf, L., Bevalot, F., Boyer, B., Fanton, L., Gaillard, Y., 2014. Screening approach by ultra-high performance liquid chromatography-tandem mass spectrometry for the blood quantification of thirty-four principles of plant origin: application to forensic toxicology. Journal of Chromatography B 975, 65–76.
Cutler, R., 2003. The Mysterious Death of Jane Stanford. Stanford General Books.
Gallagher, S., August 10, 2004. Andre Noble. Filmmaker Blog. http://www.filmmakermagazine.com/blog/2004/08/andre-noble.php/#.VQtKzWYWFFU.
Gallop, D., 2009. Plato's *Phaedo*. Oxford University Press, Oxford.
Holland, J.G., 1866. The Life of Abraham Lincoln. Gurdon Bill, Springfield, IL.
Jhala, C.I., Jhala, K.N., 2012. The Hippocratic oath: a critical analysis of the ancient text's relevance to American and Indian modern medicine. Indian Journal of Pathology and Microbiology 55, 279–282.
Levine, M., Ruha, A., Graeme, K., Brooks, D., Canning, J., Curry, S., 2011. Toxicity in the ICU part 3: natural toxins. Chest 140, 1357–1370.
Lewis, W.H., Elwin-Lewis, M.P.F., 2003. Medical Botany: Plants Affecting Human Health. John Wiley & Sons, New York.
Magner, L.N., 1992. A History of Medicine. Marcel Dekker Inc., New York.
Marty, F., 1999. Plant vacuoles. The Plant Cell 11, 587–599.
Mauseth, J., 2012. Botany: Introduction to Plant Biology, fifth ed. Jones and Bartlett Learning.

Miller, R.B., 1994. Identification of wood fragments in trace evidence. In: Proceedings of the International Symposium on the Forensic Aspects of Trace Evidence. U.S. Department of Justice, Federal Bureau of Investigation, Quantico, VA, pp. 91–111.

Morris, S.C., Lee, T.H., 1984. The toxicity and teratogenicity of *Solanaceae* glycoalkaloids particularly those of the potato *(Solanum tuberosum)*: a review. Food Technology in Australia 36, 118–124.

PBS 31 July 2013. Who killed Lindbergh's baby? http://www.pbs.org/wgbh/nova/tech/killed-lindbergh-baby.html.

Phillips, G., 2004. Alexander the Great. Murder in Babylon. Virgin Books.

Raven, P.H., Evert, R., Eichhorn, S.E., 2012. Biology of Plants, eighth ed. W.H. Freeman.

Simoons, F.J., 1998. Plants of Life, Plants of Death. University of Wisconsin Press, Madison, WI.

Simpson, B.B., Ogorzaly, M.C., 1995. Plants in Our World. McGraw Hill, New York.

CHAPTER

2

Suitability of Forensic Plant Science Evidence for Courtroom Presentations

Forensic science has become an essential investigative tool in modern crime investigations. It often determines who is/are the suspect/s and leads to arrests. If the suspect confesses or plea bargains to a lessor charge, the quality of the forensic science leading to their arrest may never be challenged. However, when forensic science is brought into the courtroom as evidence, it is a very different story.

In 2009, the National Academies of Sciences released the **NAS Report** (National Research Council of the National Academies, 2009) essentially damming the state of forensic science in the United States. According to the NAS Report, the state of forensic science is a threat to accurate investigation and innocent defendants. It proposes a series of 10 recommendations to improve forensic science in the USA. These recommendations are summarized by Risinger (2010; see Table 2.1).

1. THE CURRENT STATE OF FORENSIC SCIENCE IN THE USA

Challenges to modern forensic science in the courtroom come from two main sources. The first is from perceptions of forensic science by the general public and the second is from the scientific community (i.e., the NAS Report).

1.1 Public Perception Problems: The "Crime Scene Investigation Effect"

The prevalence of depicting fictionalized versions of real forensic science on television crime scene investigation (CSI) programs, the general public has developed an unrealistic impression of what can be done forensically and how rapidly it can be done. Consequently, juries may expect the prosecution to provide evidence using techniques they believe are real but which may have been embellished or entirely fabricated by TV writers. Although some claim that these perceptions do not alter jury decisions (e.g., Sheldon, 2008), others provide examples where the outcome of actual cases has been affected (e.g., Willings, 2004; as cited

TABLE 2.1 Summary of Recommendations from the NAS Report[a]

First and foremost, a new independent federal agency (the National Institute of Forensic Science, or NIFS) should be established and charged with authority to establish and enforce best practices for forensic science laboratories and professionals. This would include authority to establish standards for accreditation and certification, and also authority to promote necessary and appropriate research. NIFS should fund research to determine the accuracy and reliability of those currently used techniques that lack data on these issues, and such research should examine those techniques across the conditions that present themselves in practice.
As far as laboratory organization is concerned, all forensic science laboratories should be removed from the administrative control of law enforcement agencies.
NIFS should make sure that all work in forensic laboratories is properly documented using standard procedures and terminology, and that all resulting testimony is clear and uses standard forms of expression calculated to communicate the true meaning of the results of various forensic assays. There should also be in place in every lab a set of quality control procedures designed to identify mistakes, fraud, and bias and to ensure that best practices are followed.
Accreditation of laboratories and individual certification of practitioners should be mandatory. There should also be a standard code of ethics for all forensic science practitioners with enforcement mechanisms.
NIFS should fund research on the effects of observer bias to determine whether it currently affects the results of forensic examinations and, if so, how much.
NIFS should provide money to underwrite both academic training of forensic science personnel and the development of a normal research infrastructure in the academy.

[a] *Reprinted with permission from Risinger (2010).*

by Stevens, 2008). Some conclude that the "**CSI effect**" is very real and is influencing every aspect of the criminal justice system (see Dural, 2010).

1.2 Scientific Problems with Modern Forensic Science

Forensic crime laboratories in the USA are under attack as a result of the discovery of error and falsification of data, as well as from the NAS Report. Although documentation of faulty investigation and prosecutions has resulted in miscarriages of justice, wrongful convictions also have resulted from poor forensic science. A National Institute of Justice panel study directed by Jon Gould (2013) (http://www.prweb.com/releases/2013/3/prweb10513834.htm) identified the 10 most common causes for wrongful convictions (see Table 2.2).

The widespread availability of highly sensitive DNA testing has been responsible for reversals of many convictions (e.g., see The Innocence Project). Numerous DNA-reversed convictions involved erroneous eyewitness accounts (not specifically indicated in the Gould panel listing, Table 2.2), but others involved cases with questionable forensic laboratory work. Examples of shoddy work and deliberate falsification of laboratory results from forensic labs around the country have been brought to light in recent years (Hansen, 2013), leading to reevaluation of many convictions. In response to widespread questioning of forensic results, the NAS Report found that there was too much emphasis on "forensic" and not sufficient "science" in actual practice, especially in some disciplines. Consequently, the U.S. Departments of Justice and Commerce established a **National Commission on Forensic Science** to make recommendations for strengthening forensic sciences (U.S. Department of Justice, 2014).

TABLE 2.2 Ten Reasons for Wrongful Convictions[a]

State death penalty culture/state punitiveness
Strength of prosecution's case
Prosecution withheld evidence (Brady violation)
Forensic evidence errors
Strength of defendant's case
Age of defendant
Criminal history of defendant
Intentional misidentification
Lying by noneyewitness
Family witness testified on behalf of defendant

[a] *Based on Gould (2013).*

Pessimism has been expressed by some who believe nothing will really change in forensic science because Congress will not adequately fund the needed revolution and/or that Congress will create a bureaucracy that will only impede progress. Additionally, it is suggested that some forensic groups (e.g., the Federal Bureau of Investigation (FBI), the International Association of Investigators (IAI), the Scientific Working Group on Friction Ridge Analysis, Study and Technology (SWGFAST), and the Association of Firearm and Tool Mark Examiners (AFTME)) already "believe" there is no need for extensive revision of their procedures in the areas of question (Cole, 2010; Gabel, 2014). Others suggest that rather than the problem being due to faulty forensic science, many of these wrongful convictions are a consequence of overzealous prosecution of otherwise weak cases built on circumstantial evidence or evidence that was overlooked due to confirmation bias (Collins, 2015). Hence, perhaps forensic science already has enough checks and balances, at least in some areas.

2. COURT DECISIONS CONCERNING PRESENTATION OF SCIENTIFIC EVIDENCE AND EXPERT OPINION

Forensic science is an essential tool for catching criminals, but the science must be acceptable in the court system as well. However, even the best forensic science relies on the lawyers and the judiciary to do their jobs in court, too.

The Frye Standard for scientific evidence was introduced in 1923 (Frye vs US) and was considered the sole criterion for acceptability in the courtroom until 1993. Essentially, the evidence was admissible if the scientific methods employed to obtain the data were generally accepted by most researchers in that particular area (e.g., toxicology, analytical chemistry, etc.). But in addition to providing data, the forensic scientist is often asked to provide his/her expert opinion in relation to data they have collected or data collected by other scientists and to perhaps extrapolate to other situations. How does expert opinion differ from "popular opinion" that we all develop and espouse? Popular opinion may in fact have no logical relationship to the truth. One dictionary defines opinion as "a belief or conclusion held with confidence but not substantiated by positive knowledge or proof" (http://www.thefreedictionary.com/opinion). Popular opinion is subject to irrational pressures, special interests, and seductive influences of all kinds. It is often very passionate and unaffected by factual information. We see this commonly in the statements of some celebrities and politicians as well as in opinions expressed on TV and radio talk shows. In contrast, expert opinions involve rational inferences and interpretations based on experience with validated scientific data.

In 1975, **Rule 702 of the Federal Rules of Evidence** attempted to clarify the nature of expert opinion: "If scientific, technical, or other specialized knowledge will assist the trier of fact to understand the evidence or to determine a fact in issue, a witness qualified as an expert by knowledge, skill, experience, training, or education may testify thereto in the form of an opinion or otherwise." Today, the current standard for expert testimony results from three important court cases (Table 2.3): **Daubert versus Merrill Dow Pharmaceuticals, Inc.** (1993), **General Electric versus Joiner** (1997), and **Kumho Tire Ltd versus Carmichael** (1999) (see Risinger et al., 2002; Houck and Siegel, 2006).

The Dauberts gave birth to a deformed child and claimed the cause was a drug, Bendectin®, that the mother had taken during pregnancy. They submitted expert evidence suggesting that Bendectin® could cause birth defects. However, this evidence was based on in vitro and in vivo animal studies and some pharmacological studies using methodologies that had not yet gained acceptance within the general scientific community. The **Daubert Decision** stated that testimony must (1) be testable and have been tested through the scientific method, (2) have been subject to peer review, (3) have established standards, (4) have a known or potential error rate, and (5) have widespread acceptance by the relevant scientific group. Consequently, the Dauberts' evidence was excluded.

Joiner claimed his lung cancer was a result of exposure to polychlorinated biphenyls (PCBs) produced by General Electric. His claim was based on scientific studies conducted on infant mice. The decision for this case noted that these mice developed a different form of lung cancer than Joiner expressed and that in studies of adult mice, PCBs did not cause lung cancer. Therefore, the court ruled that those data could not be admitted as evidence to support Joiner's claim.

Carmichael sued **Kumho** Tire Ltd after a tire on a minivan had exploded causing the minivan to crash and injuring or killing the occupants. The testimony of the Carmichael family's tire expert was excluded because he failed to employ the same standards used by similar experts. This decision extended the consideration of testimony based on hard data to expert opinion related to skill or experience-based observations. Essentially, the Kumho decision concluded that (1) expert witnesses can develop theories based on their observations and experience and apply those theories to the case before the court, (2) all forms of expert witness testimony should be evaluated with the same level of rigor, and (3) the Daubert standards are flexible *guidelines* that may not be applicable in every instance of expert testimony.

Thus the resultant "**Daubert trilogy**" modifies Rule 702 (see Table 2.3). It essentially states that (1) testimony must be based upon sufficient facts or data, (2) testimony must be the product of reliable principles and methods, and (3) the witness must apply the principles and methods reliably to the facts of the case.

TABLE 2.3 Criteria Used for Acceptance of Forensic Evidence

FRYE versus United States (1923)[a]
Science must be generally accepted in the relevant scientific community "…the thing from which the deduction is made must be sufficiently established to have gained general acceptance in the particular field to which it belongs."
Rule 702 Testimony by Expert Witnesses (1975)[b]
"If scientific, technical, or other specialized knowledge will assist the trier of fact to understand the evidence or to determine a fact in issue, a witness qualified as an expert by knowledge, skill, experience, training, or education may testify thereto in the form of an opinion or otherwise."
Rule 702 (Amended 2000, 2011)[c]
A witness who is qualified as an expert by knowledge, skill, experience, training, or education may testify in the form of an opinion or otherwise if the: **1.** Expert's scientific, technical, or other specialized knowledge will help the trier of fact to understand the evidence or to determine a fact in issue; **2.** Testimony is based on sufficient facts or data; **3.** Testimony is the product of reliable principles and methods; and **4.** Expert has reliably applied the principles and methods to the facts of the case.
Daubert versus Merrell Dow Pharmaceuticals (1993)[d]
"trial judge must ensure that any and all scientific testimony or evidence admitted is not only relevant, but reliable." "evidentiary reliability will be based upon scientific validity." Content of testimony must **1.** Be testable and have been tested using scientific methods **2.** Been subject to peer review **3.** Have established standards **4.** Have a known or potential error rate **5.** Have widespread acceptance by relevant science group

[a] *Houck, H.M. and Siegel, J.A., 2006.* Fundamentals of Forensic Science. *Elsevier Academic Press.*
[b] *Risinger, M.D., Daks, M.J., Thompson, W.C., Rosenthal, R., 2002. The Daubert/Kumho implications of observer effects in forensic science: hidden problems of expectation and suggestion.* California Law Review *90, 1–56.*
[c] *Cornell Law School. http://www.law.cornell.edu/rules/fre/rule_702.*
[d] *USDOJ, 2014. http://www.justice.gov/opa/pr/2013/February/13-dag-203.html.*

It becomes important for judges, prosecutors, defense lawyers, and jurors to distinguish assumptions made by experts versus reasonable generalizations based on scientific facts. Hence, they need to filter testimony and detect if expert opinion is clouded by bias and/or common opinion; that is, what conclusions can an expert witness draw from a data set and when is the witness simply speculating. Often, scientific experts must explain how their methodology bridges the gap between the evidence and their conclusions. In other words, the expert must justify his/her conclusions. Ultimately, it is up to the judge to accept or reject the testimony.

Typically, this is done in a separate hearing, sometimes called a **Daubert inquiry**, where no jury is present. Failure to conduct such a hearing has resulted in reversals by higher courts (http://www.ims-expertservices.com/bullseye-blog/december-2012/no-daubert-hearing-equals-$10-million-error-in-9th-circuit/).

It is questionable, however, that judges actually have the expertise to evaluate scientific evidence or to be knowledgeable enough to even recognize who are the relevant experts. Some scientific fields are easier to evaluate because the data may speak for themselves and extrapolations by the experts are obviously reasonable. Toxicological analysis, for example, is based on the general principles and methodologies of chemistry and physiology that have been validated through decades of careful study. It is relatively easy for the forensic toxicologist to establish the reliability of the methods employed and the accuracy of results obtained. Similarly, the identification of plant species via anatomical and morphological features has been developed and firmly established through centuries of careful, repeatable work. The chemical identification procedures of toxicologists employ the determination of unknown chemicals that are compared to known standard chemicals. Similarly, the plant scientist compares unknown tissues of plants as well as plant fragments or whole plants to known tissues or plants. The criteria used are clear and unambiguous, and judges or juries readily can observe and assess the results.

In contrast, several other forensic areas requiring conclusions and opinions by forensic scientists have recently come under attack. Two of these targeted forensic areas, where obvious errors in identification have been discovered are fingerprinting (i.e., friction ridge skin patterns; e.g., see McMurtrie, 2010; Dror and Mnookin, 2010; also http://en.wikipedia.org/wiki/Brandon_Mayfield) and bite mark identification (e.g., see Rix, 2007; Balko, 2015a,b,c,d). These areas reportedly require a high degree of subjective observation, which has led to considerable disagreement about the reliability of conclusions by these forensic experts. For example, critics claim that reproducibility and reliability have not been established by rigorous testing of friction ridge skin patterns (Haber and Haber, 2008; Cole, 2010; Dror and Mnookin, 2010; Ulrey et al., 2014; Mustonen et al., 2015), and that there can be considerable disagreement among fingerprint experts tested with the same materials. However, a study supported by the National Institutes of Justice was recently conducted using a standardized testing paradigm on 109 fingerprint examiners from 76 federal, state, and local agencies across the USA (Pacheco et al., 2014). They report a very low error rate in matching prints using their testing protocol. This report has not been peer reviewed, however, and it awaits evaluations by critics of the current state of the art.

There is an inherent bias in the Daubert criteria that is a disadvantage to the opposite side where forensic evidence can be introduced without having the forensic scientist/technician present for cross-examination. If the judge allows it as evidence, there is no confrontation of the expert witness possible. The Melendez-Diaz versus Massachusetts decision stated that "testimonial evidence" violates a defendant's constitutional right to confrontation (see Moreno, 2010).

2.1 What Criteria Determine Validity?

It is important to realize that science in general is struggling with the concepts of statistical significance and repeatability. This is not just a concern for forensic sciences. It has been the custom for decades to assume that a statistical test that indicates a **probability (*p*)** of less than 0.05 means the data and associated conclusions about the data are valid (i.e., significant). If $p<0.01$, then it is "very significant." This is generally interpreted to mean that if the same study were repeated, at least 95 or 99% of the time, respectively, one would obtain the same results.

The lower the *p* value, the more likely the results of an experiment will be accepted for publication. Unfortunately, this search for a significant *p* has sometimes led to what has been termed "**data mining**"; i.e., if $p > 0.05$, the investigator searches for a different test that will provide "significance" or the investigator may even alter the data set by eliminating the so-called "outliers" (see http://en.wikipedia.org/wiki/Outlier). If two different statistical tests of the data yield opposite results, which one should be retained? Might confirmation bias play a role here (see ahead)?

Statisticians are currently suggesting that reliance on $p < 0.05$ is a misinterpretation of what *p* represents (Nuzzo, 2014). Rather, *p* is the probability of whether or not the **null hypothesis**[1] would have been supported by the data, not a measure of reproducibility as generally assumed.

2.2 Objectivity in Forensic Analyses Is of Paramount Importance

Forensic scientists and technicians must be ever vigilant to retain their objectivity and not be influenced by **observer effects** (Table 2.4; Risinger et al., 2002). Whenever possible the forensic investigator should be isolated from the sociological details of the crime (especially personal details about the suspect) as well as the opinions of the crime scene investigators or prosecutors. The simple act of requesting an analysis can affect the objectivity of a forensic scientist (Whitman and Koppl, 2010).

Opinions of law enforcement personnel or superiors can result in **conformity effects** altering the objectivity of the forensic investigator. Even the opinions of individuals held in low esteem can affect one's objectivity. Thus, the opinion of an inexperienced person may be discounted without thorough consideration. Such biases are results of **anchoring effects**.

One of the most dangerous observer effects is **confirmation bias**, the tendency to look for instances that confirm your own hypothesis (see also Byrd, 2006). A forensic scientist may find reasons to keep repeating tests or try new approaches to achieve the results that support her/his hypothesis instead of accepting the data already obtained. Replication is a hallmark of good science, but unnecessary repetition may simply waste valuable time and resources. Forensic scientists are no more likely to fall into this trap than other research scientists as it is often difficult to accept that your favorite hypothesis is incorrect. Observer effects can lead to production of errors throughout the forensic process (see Table 2.5).

Some suggest it is not possible to eliminate biases even if forensic scientists were perfectly rational in reaching their conclusions (Whitman and Koppl, 2010). Forensic scientists necessarily make many subjective evaluations in the process of examining evidence. Furthermore, only one crime lab typically is involved in an analysis and the same lab will be responsible for the interpretation of the data they generated.

[1] The null hypothesis is a statistical construct that states there is no difference between the two (or more) groups being compared. This is generally not what a scientist typically does when he/she formulates a hypothesis to be tested experimentally where they suggest, for example, that a certain treatment will produce a certain effect.

TABLE 2.4 Some Observer Effects in Forensic Science

1. Random error effects	Random errors made by observer; not due to bias.
2. Confirmation bias effects	Observer sees what he/she expects or wants to see. Can also affect decision thresholds causing false positives or negatives (Phillips et al., 2001).
3. Conformity effects	Tendency to conform to perceptions, beliefs, and behavior of others (Risinger et al., 2002); especially common with responses to supervisors, heros, or experts.
4. Anchoring effects	Bias induced by external information leading to subjective responses (Mussweiller and Stack, 1999).
5. Role effects	Once assigned a role, an observer views/remembers data differently from an observer assigned a different role. For example, a forensic worker may adopt the role of the prosecution and lose objectivity as they attempt to bring a suspect to "justice" (Starrs, 1971).

TABLE 2.5 Errors Resulting From Observer Effects[a]

Stage	Description
Errors of apprehending	Errors of initial perception
Errors of recording	Errors made during initial observation assuming a written record is kept (may include random errors unrelated to observer effects)
Errors of memory	Errors induced by desires and the need for consistency (especially important when there is no written record or only sketchy notes)
Errors of computation	Errors that may occur as a result of transformation of data using incorrect methods or simply random errors
Errors of interpretation	Errors made in drawing conclusions

[a] *Risinger, M.D., Daks, M.J., Thompson, W.C., Rosenthal, R., 2002. The Daubert/Kumho implications of observer effects in forensic science: hidden problems of expectation and suggestion. California Law Review 90, 1–56.*

2.3 Repeatability

Good science relies on the concept that repetition of a well-designed experiment will yield the same result. **Repeatability** is considered a hallmark of good science, although the present culture of publication and grant support discourages scientists from repeating other peoples' studies. Instead, scientists rely heavily, and sometimes blindly, upon statistical analyses of data to determine the significance of data sets although these analyses are not infallible and are often misinterpreted (see above). Forensic treatments that cannot provide statistical support and/or evidence of error rates may be open to criticism. However, some observational science does not lend itself to standard statistical tests, yet these analyses may be as valid as those that do.

2.4 How Is the Forensic Community Responding?

As expected, the NAS Report has met with mixed reactions. The history of attempts at political modification of forensic science suggests that although some improvements may result, the problems outlined in their report will not be solved overnight (see Gabel, 2014). Some seem to believe that accreditation of national forensic laboratories will solve the problems and this will trickle down to state laboratories. Others point out that accreditation requires development of standards and will require considerable research before accreditation will be possible. Removal of prosecutorial bias in forensic laboratories will require the separation of laboratories physically and financially from law enforcement and public safety agencies. An analysis of 300 forensic laboratories showed that 79% were linked geographically and/or financially to law enforcement or public safety agencies (Giannelli, 2010). Formation of alliances between research units at universities with forensic laboratories doing the applied work has been proposed as a more reasonable way to improve forensic science with respect to the NAS guidelines for improvement without creating a huge bureaucracy (Gabel, 2014).

Some are pessimistic that major organizations responsible for forensic information will change in light of the NAS Report (Cole, 2010; Koehler, 2010). Some insist that in spite of documented errors made, the forensic approaches they use have considerable historical basis and that is enough to validate them. Clearly, the NAS Report points out that more standardization of procedures and testing of the accuracy of these more subjective techniques involving judgment calls based on experience are needed (e.g., in the case of friction ridge analyses, bite mark evidence, tool markings, bloodstain patterns, etc.).

Training procedures for forensic scientists also must be reexamined. One good sign is the recent establishment of the Forensic Science Institute (FSI) at the University of Central Oklahoma. FSI provides undergraduate and graduate training in the forensic sciences, including a laboratory directed at digital and cybercrime (Adams et al., 2013). Hopefully, that training will emphasize a better understanding of validation and replication that will approach the standards outlined in the NAS report.

Finally, the lack of adequate financial support for federal, state, and local crime laboratories has blocked historical efforts at improving forensic science (Gabel, 2014). Many doubt that the U.S. Congress will ever provide the necessary funds for such an endeavor and that all the efforts embodied in the NAS Report will be for naught. Only time will tell.

3. HOW DOES DAUBERT RELATE TO FORENSIC PLANT SCIENCE

Forensic plant science has been accomplished almost exclusively by scientists who reside in academic institutions. These scientists have obtained masters and PhD degrees in botany, biology, environmental science, etc. They have received training in one or more special aspects of plant science (anatomy, genetics, morphology, systematics, taxonomy, plant ecology, palynology, algology, diatomology, etc.). They often are willing to provide their expertise to law enforcement personnel but they usually are not trained in forensic science. Therefore, it becomes the task of the investigator to inform the plant scientist of standard forensic procedures, how to handle evidence, etc. until they have become familiar with the routines.

The forensic plant scientist primarily uses procedures and tools that for the most part have been used for centuries (see Chapters 4, 6, 8, and 10). Plant identification (taxonomy) uses anatomical and morphological clues to compare known plants with unknown plants.

Plant ecology relies in part on taxonomy and in part on sound ecological principles that are accepted among ecologists worldwide. Consequently, it is relatively easy for plant scientists to fulfill the Daubert requirements.

The uses for DNA analyses so far have been limited for forensic plant science due to the huge numbers of known plant species, the few genomes that have been sequenced, and the relative ease of identification through anatomical and morphological features (see Chapter 3). Genomes sequenced to date cover mostly economically important plants. New techniques have been developed for identifying some species but these approaches have not been validated as those for comparison of mitochondrial and nuclear DNA has been for humans. Using DNA technology, one cannot yet reliably separate individuals of the same species by any cost-effective method, but this may change as new technologies are developed.

To date, there is no official program for training people in forensic plant science and it is highly unlikely that most forensic laboratories could afford to maintain a fulltime plant scientist. Furthermore, there is no certification system in place for forensic plant scientists. Consequently, qualifications as an expert in forensic plant science must be done on an individual basis for each scientist. However, many of the procedures we describe in this book can be readily learned and validated by any forensic expert. Hopefully, such basic training in the plant sciences will become routine in forensic training centers in the near future. Meanwhile, the use of trained plant scientists at universities and other research centers can fill the void.

References

Adams, D.E., Mabry, J.P., McCoy, M.R., Lord, W.D., 2013. Challenges for forensic science: new demands in today's world. Australian Journal of Forensic Science 45, 347–355.

Balko, R., 2015a. How the Flawed 'Science' of Bite Mark Analysis Has Sent Innocent People to Jail. Washington Post. 13 February. http://www.washingtonpost.com/news/the-watch/wp/2015/02/13/how-the-flawed-science-of-bite-mark-analysis-has-sent-innocent-people-to-jail/.

Balko, R., 2015b. It Literally Started with a Witch Hunt: A History of Bite Mark Evidence. Washington Post. 17 February. http://www.washingtonpost.com/news/the-watch/wp/2015/02/17/it-literally-started-with-a-witch-hunt-a-history-of-bite-mark-evidence/.

Balko, R., 2015c. Attack of the Bite Matchers. Washington Post. 18 February. http://www.washingtonpost.com/news/the-watch/wp/2015/02/18/attack-of-the-bite-mark-matchers-2/.

Balko, R., 2015d. The Path Forward on Bite Mark Matching – and the Rearview Mirror. Washington Post. 20 February. http://www.washingtonpost.com/news/the-watch/wp/2015/02/20/the-path-forward-on-bite-mark-matching-and-the-rearview-mirror/.

Byrd, J.S., 2006. Confirmation bias, ethics, and mistakes in forensics. Journal of Forensic Investigation 56, 511–525.

Cole, S.A., 2010. Who speaks for science? A response to the National Academy of Sciences report on forensic science. Law, Probability and Risk 9, 25–46.

Collins, J., 2015. http://www.science20.com/keeping_the_gate_a_science_and_justice_blog/blog/the_law_has_failed_not_forensic_science-153589.

Dror, I.E., Mnookin, J.L., 2010. The use of technology in human expert domains: challenges and risks arising from the use of automated fingerprint identification systems in forensic science. Law, Probability and Risk 9, 47–67.

Dural, E.W., 2010. Crime scene investigation (as seen on TV). Forensic Science International 199, 1–5.

Gabel, J.D., 2014. Realizing reliability in forensic science from the ground up. Journal of Criminal Law & Criminology. 104, 283–352. http://scholarlycommons.law.northwestern.edu/jclc/vol104/iss2/2.

Giannelli, P.C., 2010. Independent crime laboratories: the problem of motivation and cognitive bias. Utah Law Review 2010, 247–266.

Gould, J., 2013. http://www.prweb.com/releases/2013/3/prweb10513834.htm.

Haber, L., Haber, R.N., 2008. Scientific validation of fingerprint evidence under Daubert. Law Probability and Risk 7, 87–109.

Hansen, M., September 2013. Crime labs under the microscope after a string of shoddy, suspect and fraudulent results. ABA Journal 99 (9). http://www.abajournal.com/magazine/article/crime_labs_under_the_microscope_after_a_string_of_shoddy_suspect_and_fraudu/.

Houck, H.M., Siegel, J.A., 2006. Fundamentals of Forensic Science. Elsevier Academic Press.

Koehler, J.J., 2010. Forensic science reform in the 21st century: a major conference, a blockbuster report and reasons to be pessimistic. Law, Probability and Risk 9, 1–6.

McMurtrie, J., 2010. Swirls and whorls: litigating post-conviction claims of fingerprint misidentification after the NAS report. Utah Law Review 2010, 267–296.

Moreno, J.A., 2010. C.S.I. bulls#!t: the National Academy of Sciences, Melendez-Diaz v. Massachusetts, and future challenges to forensic science and forensic experts. Utah Law Review 2010, 327–365.

Mussweiler, T., Strack, F., 1999. Comparing is believing: a selective accessibility model of judgmental anchoring. European Review of Social Psychology 10, 215.

Mustonen, V., Kakkarainen, K., Tuuneinen, J., Pahjola, P., 2015. Discrepancies in expert decision-making in forensic fingerprint examination. Forensic Science International 254, 215–226.

National Research Council of the National Academies, 2009. Strengthening Forensic Science in the United States: A Path Forward. National Academies Press, Washington, DC.

Nuzzo, R., 2014. Statistical errors. Nature 506, 150–152.

Pacheco, I., Cerchiai, B., Stoiloff, S., 2014. Miami-Dade Research Study for the Reliability of the ACE-v Process: Accuracy & Precision in Latent Fingerprint Examinations. National Institutes of Justice. Document No. 248534.

Phillips, V.L., Royer, J.M., Greene, B.A., 2001. The application of signal detection theory to decision-making in forensic science. Journal of Forensic Science 46, 294–296.

Risinger, M.D., 2010. The NAS/NRC report on forensic science: a path forward fraught with pitfalls. Utah Law Review 2010, 246–255.

Risinger, M.D., Daks, M.J., Thompson, W.C., Rosenthal, R., 2002. The Daubert/Kumho implications of observer effects in forensic science: hidden problems of expectation and suggestion. California Law Review 90, 1–56.

Rix, J., 2007. Jingle Jangle: The Perfect Crime Turned Inside Out. Broken Bench Press, Zephyr Cove, NV.

Sheldon, D.E., 2008. The "CIS effect": does it exist. National Institutes of Justice Journal 259, 1–6.

Starrs, J.E., 1971. The ethical obligations of the forensic scientist in the criminal justice system. Journal of the Association of Official Analytical Chemists 54, 906–910.

Ulery, B.T., Hicklin, R.A., Roberts, M.A., Buscaglia, J., 2014. Measuring what latent fingerprint examiners consider sufficient information for individualization determinations. PLoS One 9, e110179. http://dx.doi.org/10.1371/journal.pone.0110179.

U.S. Department of Justice, 2014. http://www.justice.gov/opa/pr/2013/February/13-dag-203.html.

Whitman, G., Koppl, R., 2010. Rational bias in forensic science. Law, Probability and Risk 9, 69–90.

Willing, R., 2004. CSI Effect Has Juries Wanting More Evidence. USA Today 5 August. Cited by Stevens, D.J., 2008. Forensic science, wrongful convictions, and American prosecutor discretion. The Howard Journal (of Criminal Justice) 47, 31–51.

CHAPTER

3

Sources for Forensic Plant Science Evidence

Man's capacity for justice makes democracy possible, but man's inclination to injustice makes democracy necessary.
Reinhold Niebuhr (American theologian)

In this chapter, we provide a brief overview of four approaches that are useful in forensic plant science: plant anatomy, plant taxonomy, plant ecology, and genetic analyses. We have used the first three approaches for over 30 years. These three approaches are uncomplicated, inexpensive, and relatively easy to learn. Genetic analyses are much more expensive to conduct, and require sophisticated equipment as well as specialized training. In subsequent chapters, we will treat plant anatomy (Chapter 4), plant taxonomy (Chapter 6), and plant ecology (Chapter 8) in more detail. Information concerning the forensic use of diatoms and pollen is provided in Chapter 10. Additional information on genetic analyses can be found in the references cited at the end of this chapter.

1. PLANT ANATOMY

Plant anatomy is the study of the structure of plant cells. The emphasis in our work has been on flowering plants that we consume as food. We have pioneered the use of food plant cells to help determine time of death, to corroborate or refute sworn testimony, and to connect a suspected victim to a particular site. Examination of stomach contents, intestinal contents, and even fecal materials can be used to identify plants consumed. In the case of stomach contents, knowledge of the last known meal can help fix the time of death (see Chapter 4). Fecal stains on a suspect's clothing can link him/her to a rape or rape/homicide victim (Norris and Bock, 2000). Examples of the use of plant anatomy in actual forensic investigations are provided in Chapter 5. Excellent modern plant anatomy books are authored by Katherine Esau (1977) and her students (Evert and Eichhorn, 2006). These works are technical, but the writing is clear, and motivated people will find them useful.

There are only a couple dozen general types of plant cells, many fewer than those found in animals (see Chapter 4 for details on plant cells). Plant anatomy as a scientific discipline became prominent because of two inventions: the light microscope in the late 1500s

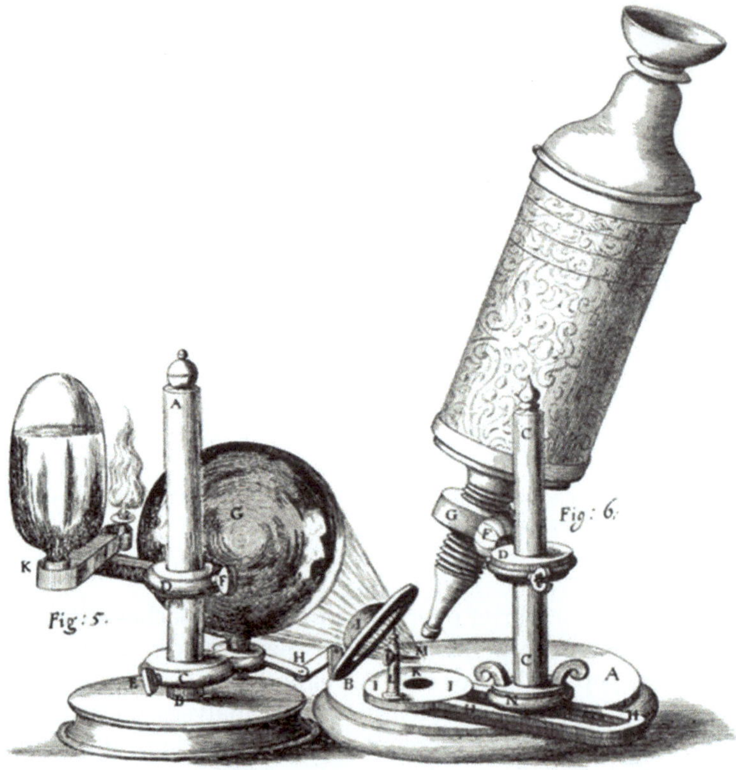

FIGURE 3.1 Hooke's microscope.

(Clay and Court, 1932) and an earlier invention around 1440, the printing press (Moran, 1973). Robert Hooke used an advanced form of the early microscope (Figure 3.1) to examine natural materials. He published his microscopic observations in the magnificent book, *Micrographia* (Hooke, 1665). In it, he showed an early, accurate picture of plant cell walls in cork oak (Figure 3.2). Two other books were instrumental in making plant anatomy into a formal branch of botany. Marcello Malpighi (1686) described his observations of plants and animals including microscopic pictures of plant cells in addition to his medical observations. Nehemiah Grew's book (1672), the first of its kind published in English, is restricted to clear pictures of plant cells.

As microscopes improved, so did plant cell depictions. The modern light microscope is sufficient for virtually all the forensic work we do in plant anatomy, and it is inexpensive. This tool is widely accessible to investigators and the general public. Most secondary schools teach the use of light microscopes in their biology classes. One major advance in microscopy came in the mid-twentieth century with the development of electron microscopy (Ruska, 1986). The subsequent development of the **scanning electron microscope (SEM)** provided very accurate and rather attractive three-dimensional photographs (Figure 3.3; see also Ledbetter and Porter, 1964), although it is more expensive than light microscopy. Furthermore, SEM microscopes are not as readily available as light microscopes. This long and

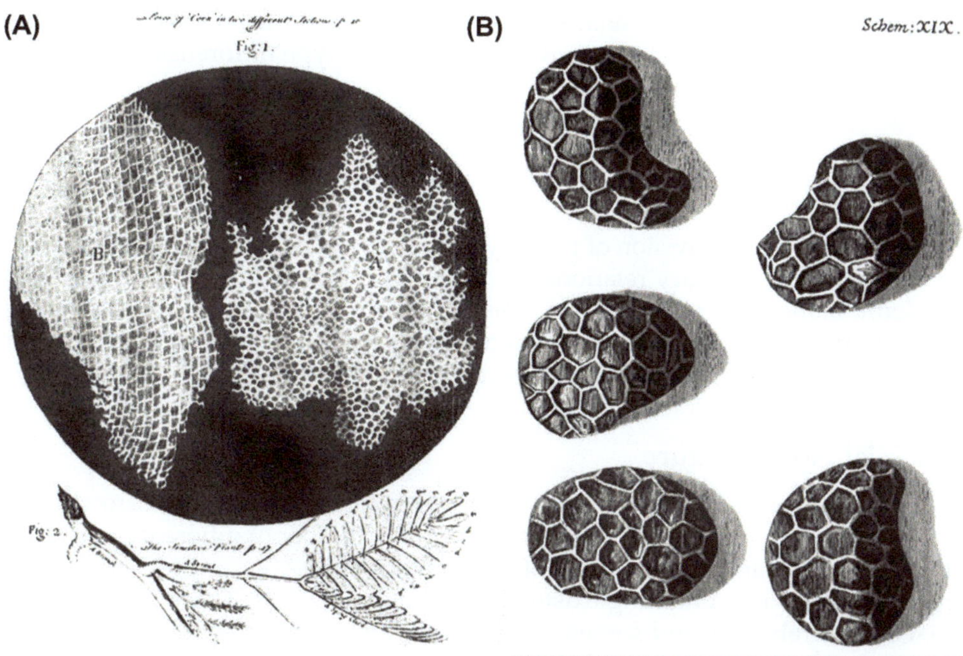

FIGURE 3.2 Early observations by Hooke. (A) This early observation of cork cell walls was the basis for the eventual discovery that plants and animals were made up from collections of cells. (B) Drawings of poppy seeds by Hooke.

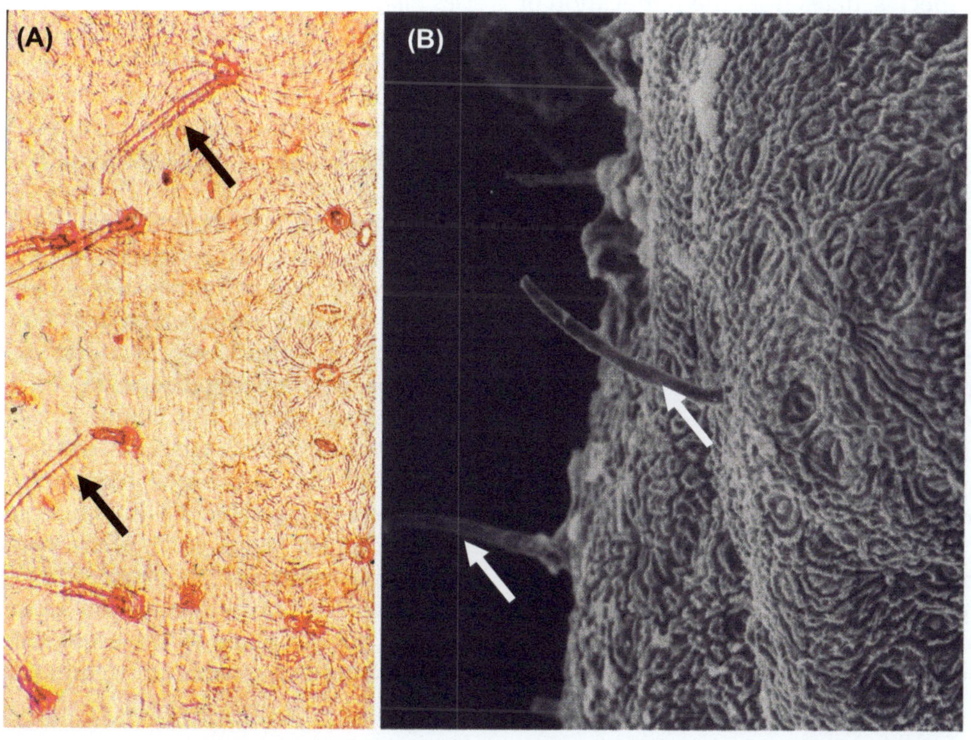

FIGURE 3.3 Comparison of the surface of a green bean as seen through (A) the light microscope and (B) the scanning electron microscope. Arrows indicate trichomes or spines as seen by using each tool. *SEM courtesy of Dr Meredith A. Lane.*

well-documented history of plant anatomy together with the time-honored approach of visually comparing unknown cells to known cells makes this type of forensic plant science evidence readily accepted in courts.

2. PLANT TAXONOMY

Plant taxonomy is a subdivision of plant systematics. The overall goal of plant systematics is to show plants' evolutionary relationships. Plant taxonomy's goal is to assign the proper scientific Latin name for a plant species. This procedure often is required in forensic work. Identification of a particular species is extremely important in cases involving poisons and illicit drugs. Plant ecology often requires species identification as well (see ahead).

2.1 Binomial Nomenclature

In order for a forensic plant scientist to communicate effectively, the correct name of the plant must be entered in evidence. Because of the great number of different plant species, the correct name requires an understanding of the naming process.

Many systems for plant naming existed before the mid-1700s. These earlier names were in Latin, Greek, Arabic, or/and sometimes in local vernacular names (Pavord, 2005). These names sometimes were several words in length. This led to the same plant having different names in different places and made communication among plant scientists difficult. The current system of **binomial nomenclature** involves identifying plants with two names, a **genus** and a **species**, dates from the publication of *Species Plantarum* in 1753 by Carl Linné, a Swedish botanist. His work published under his Latin name, Carolus Linnaeus (1753), is acknowledged as the starting place for current naming procedures (Blunt, 1971). Linnaeus had an international team of botanists gathering plants from around the world. His workers came from many countries, including Britain, France, Germany, and the Americas. Linnaeus, with broad international cooperation, served as a clearing house for plant collections. He spent most of his time assigning names to the species.

Correct names today are governed by the International Code of Botanical Nomenclature (McNeill et al., 2012; Bock, 1973). The technical name of a plant consists of two parts, a genus (pl. genera) name and a **specific epithet** that together comprise the species name. For instance, the correct name of the common dandelion is *Taraxacum officinale* where *Taraxacum* is the genus and *officinale*, the species. Especially in depositions and even in courtroom testimony, one needs to communicate exactly what plant species are being discussed.

Mammals and birds have officially sanctioned common names, but common names of plants without accompanying photos are not substitutes for their scientific names. There are approximately 10,000 species of birds and about half that number of mammals. There are over 300,000 species of flowering plants making general agreement with multiple common names inappropriate. Hence, plant scientists depend on genus and species names for proper communication. This allows people everywhere to communicate about plant species with certainty. For example, if the plant scientist is working where Russian is the spoken language and the alphabet is Cyrillic, the Latin genus and species name still is used. This rigor in naming eliminates confusion in forensic work. For example, the name for opium poppy is

FIGURE 3.4 Plant press. *Photo courtesy Forestry Suppliers, Inc. Used by permission 580101.*

Papaver somniferum. The genus *Papaver* by itself does not indicate you are talking about the opium producer because many poppy species exist in this genus, and each has its own species name.

2.2 Collection Methods for Taxonomic Evidence

Plants should be collected in paper or cloth bags. Nonporous containers including plastic bags and glass and metal cans are unacceptable because they encourage the growth of bacteria and fungi, thus compromising plant identification.

Professional botanists often collect and temporarily store plants in plant presses (Figure 3.4). The specimens are arranged between layers of newsprint so that the most of the plant's features are displayed as the plant dries. The plant press is held together in a slatted frame and tightened by adjustable straps. For forensic work, it is common for large samples to be collected in paper shopping bags and pressed later. Paper stamp and coin envelopes are used for tiny samples including plant fragments.

Similarly, the identification of plant species using morphological criteria (stems, roots, leaves, especially flowers, etc.) in conjunction with a dichotomous key and information as to what plant species should be present at a specific locale provides simple "yes," "no," or "perhaps" conclusions. Use of herbarium specimens (Figure 3.5) can help confirm the identification of a species. Sometimes it may be possible to identify a plant only to genus or just to family, yet that information still can be very useful.

When attempting to associate a plant or plants with a particular habitat, often it is best to employ the services of a local botanist with knowledge of the local flora. Individual species sometimes are linked to a particular habitat type (see Chapters 8 and 9).

3. PLANT ECOLOGY

Ecology is the study of living organisms and their environments. It may be the youngest of the natural sciences. But ecological approaches to problem solving are as old as *Homo sapiens*. People always have used ecological knowledge in the practical sense. Ancient peoples learned by

FIGURE 3.5 Herbarium specimens are prepared by pressing the plant flat and drying it. If properly prepared and cared for, they can last for many decades.

trial and error to find useful food and medicinal plants and to avoid toxic plants. And that information was passed from generation to generation by oral tradition.

The formal field of ecology traces its origin to the middle of the 1800s. The earliest use of the term "ecology" was in a letter sent by Henry David Thoreau in 1858 where he writes, "Mr Hoar is still in Concord, attending to Botany, Ecology, &c." (Goodland, 1975). Plant ecology was established formally early in the twentieth century. By the late 1800s, Charles Darwin's works were widely read and acknowledged by biologists, but it wasn't until around 1900 that Gregor Mendel's papers describing the fundamentals of genetic inheritance were rediscovered. This added genetic inheritance to Darwinian theory. Mendelian genetics and plant ecology matured together. Much early work in plant ecology emphasized which plant species tended to occur together thereby forming **plant communities**. The pattern of a plant community changes through time and is termed **plant succession**. Succession patterns were championed, sometimes to extremes, by American plant ecologists, especially Frederick Clements (Clements, 1913; Pound and Clements, 1900), as well as French and Swiss plant ecologists. Gleason (1926) offered other views. By the 1950s plant ecology was firmly established in Britain, France, Germany, Scandinavia, Eastern Europe, and North and Central America (Tansley, 1947). During this same time period, noteworthy ecological journals showed up in other countries including New Zealand, Australia, Japan, and India where vigorous ecological work was taking place.

Plant ecology and plant geography also are tightly linked, and vegetation mapping remains an important focus of ecological research. Also, plant physiological ecology is important (Billings, 1985). The combined roles of environment versus genetic endowment of plants were boosted by British and American plant ecologists (Tansley, 1947; Stebbins, 1950; Baker and Stebbins, 1965). The ecological role of soil science, the edaphic factor in plant ecology, was spearheaded by Russian plant ecologists. The **edaphic factor**, sometimes called the soil factor or the substrate factor, is critical in plant distributions. The most dramatic changes in vegetation often can be attributed to places where soils change dramatically. This can be due to spots that are polluted or places where the soil's parent material changes below the soil surface (Jenny, 2011). This maturation was aided by many advances in environmental monitoring devices such as remote sensing and handheld portable global positioning systems. Also recording devices that kept track of global temperature, wind, and climatic patterns allowed for broad scale comparisons and predictions.

Plant ecological science has wide application in forensic work. Examples include help in searching for missing persons and clandestine graves. A few years ago, the Federal Bureau of Investigation (FBI) circulated a picture of what appeared to be a kidnapped and bound person, lying across a log in a mountainous, rocky area. They asked for help in locating this place based upon the vegetation and abiotic background shown in the photo. Detailed examples of applications are provided in Chapters 6 and 8.

4. GENETIC ANALYSES: USE OF DEOXYRIBONUCLEIC ACID

Genetics is the study of the inheritance of genes composed largely of a large polymer known as **deoxyribonucleic acid (DNA)**. Most genes are found in the nucleus (nuclear DNA), but extranuclear DNA is found in the cytoplasm of the cell too. A basic acquaintance with the principles of genetics and a minimal understanding of the structure of DNA and its role in determining characteristics of an organism is assumed here. Only a brief overview is provided and the reader is referred for a refresher to convenient summaries available on the World Wide Web (e.g., wikipedia.org/wiki/introduction_to_genetics; wikipedia.org/wiki/DNA).

Nuclear DNA is responsible for determining all the basic characteristics of an organism. Nuclear genes are found on chromosomes located in the nucleus of the cell. Each cell typically has two complete sets of chromosomes, one inherited from the male parent and one from the female parent. During formation of **gametes** (female gametes or ova; male gametes, sperm or pollen) in sexually reproducing animals and most plants, the chromosome pairs are separated through a process called **meiosis** so that only one set consisting of one of each kind of chromosome is found in a gamete. For example, a human has 23 different pairs of chromosomes (total = 46). Thus, the adult is said to be **diploid** (*Gk, diplous,* double) by having two complete chromosome sets. Each gamete gets one complete set of 23 different chromosomes and is termed **haploid** (*Gk, haplos,* single). **Fertilization** (the fusion of the nuclei of male and female gametes) results in restoration of the original number of chromosomes (2 sets or 46 total) in a cell that will multiply and differentiate into the many cell types of a new organism. **Extranuclear DNA** is found in cytoplasmic organelles: mitochondria and chloroplasts.

4.1 Nuclear DNA

One molecule of nuclear DNA is composed of two long strands forming a double helix. Each strand consists of four nucleic acid bases: **guanine**, G; **cytosine**, C; **adenine**, A; **thymine**, T (Figure 3.6) in the form of **nucleotides** (Figure 3.7) arranged in varying linear order. The strands are linked through chemical bonds formed between certain bases: G pairs with C and A pairs with T. Thus, the two strand are said to be complementary in that if A appears in one strand it must be paired to a T in the other strand. The double helix resembles a spiral staircase with the phosphate–sugar chains forming the edges and the paired bases representing the steps (Figure 3.8). The nucleotides in certain regions of the DNA represent the coding regions of genes that direct all cells, organs, and organismal activities of a plant or animal. The sequences of these bases in a given gene may differ slightly from individual to individual within a species in predictable ways. These gene sequences are inherited from their parents. Additionally, the bulk of the DNA nucleotide sequences are termed **nongenomic DNA** (also called **repetitive DNA** or "junk" DNA) in that they don't directly determine traits of an organism. However, they often contain unique base sequence patterns and inherited repetitive sequences known as **short tandem repeats** (STRs; see Butler and Becker, 2001). In humans, these STRs have proven useful in characterizing a particular individual's DNA (a process called **DNA fingerprinting**).

FIGURE 3.6 The four nitrogenous bases from which the polymer of DNA is made. (A) Adenine; (B) Guanine; (C) Thymine; (D) Cytosine.

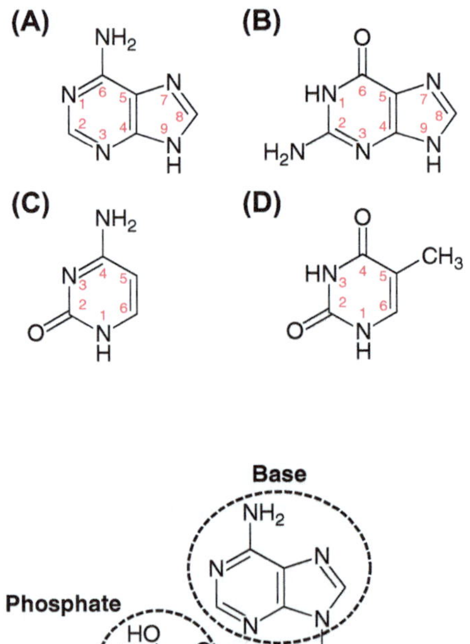

FIGURE 3.7 Nucleotide general structure. This is actually the structure of the adenine nucleotide. When nucleotides are linked together to form a single chain, they are connected through phosphate and sugar (deoxyribose) units. There are four nucleotides made from the four nitrogenous bases (see Figure 3.6) present in DNA. Two nucleotide chains are combined to form an intact double helix molecule of DNA (see Figure 3.8).

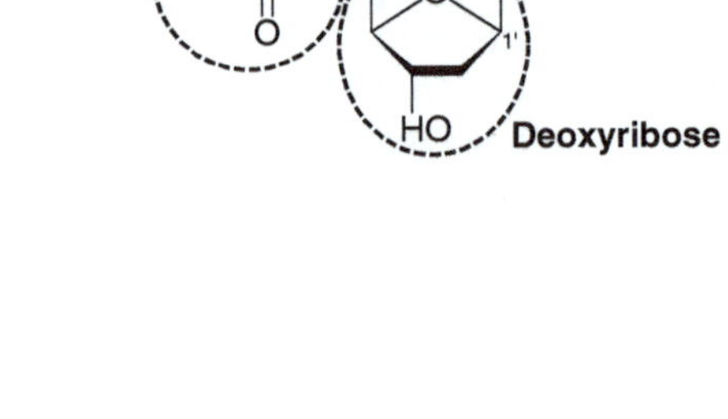

4.2 Extranuclear DNA

Extracellular DNA is found in mitochondria (**mtDNA**) of animals and plants and chloroplasts (**cpDNA**) of photosynthetic plants. mtDNA and cpDNA molecules have a circular form rather than a linear form like nuclear DNA (Figure 3.9). Since a single cell may contain many of these organelles, there are many duplicate copies of organelle DNA molecules present in a single cell.

In animals, mitochondrial genes are involved with energy production, and mitochondria are inherited only via the ova. Hence, all descendants from a single female will have essentially identical mtDNA, whereas nuclear DNA will consist of an equal mixture from both parents. This means that a female passes on her mitochondrial genes to both female and male offspring. However, there is some evidence for limited biparental inheritance in animals (Barr et al., 2005). Generally, mtDNA has been useful in human forensic studies although it does not distinguish between a mother and her children or among siblings, or from her daughter's children.

In plants, chloroplast genes are responsible primarily for photosynthesis. The pattern of mitochondrial and chloroplast inheritance in plants varies greatly. For example, both mitochondria and chloroplasts exhibit paternal inheritance in sequoia trees (Neale et al., 1989), whereas in loblolly pine mitochondrial inheritance is maternal and only chloroplast inheritance is paternal (Neale and Sederoff, 1989). Mitochondrial inheritance in other plant species may be biparental, paternal, or maternal (Birsky, 1995; Wang et al., 2010).

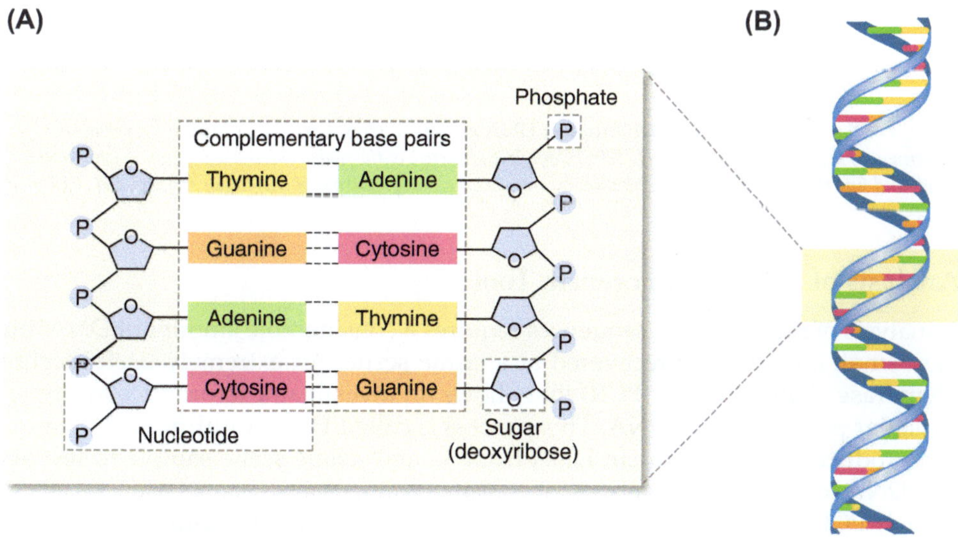

FIGURE 3.8 DNA double helix structure. (A) Complementary pairing of nucleotides with the phosphates and sugars forming the backbones and the four bases connecting in pairs. Note that only adenine pairs with thymine, whereas guanine always pairs with cytosine. (B) Schematic depicting a short segment of a DNA molecule. *Modified from bio3400.nicerweb.com.*

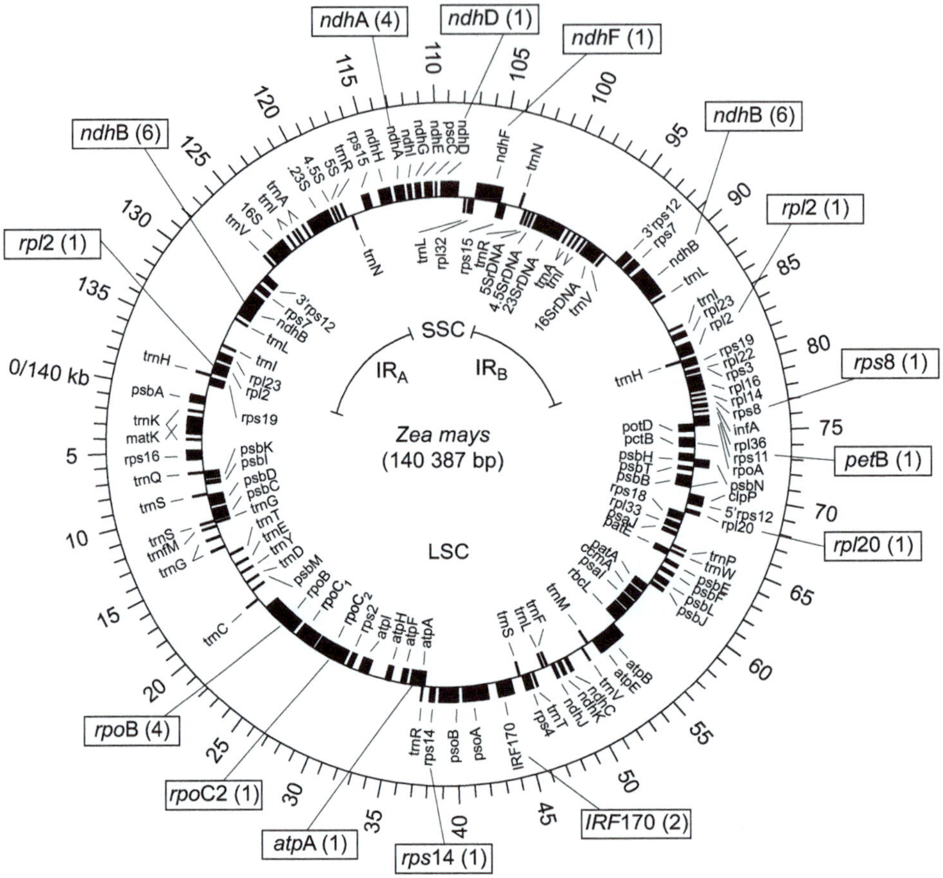

FIGURE 3.9 Plant organelle DNA. Chloroplast DNA molecule from corn. *(Reprinted with permission from Maier, R.M., Neckermann, K., Igloi, G.L., Kössel, H. 1995. Complete sequence of the maize chloroplast genome: gene content, hotspots of divergence and fine tuning of genetic information by transcript editing. Journal of Molecular Biology* **251***. 614–628.)*

4.3 Analysis of DNA as a Forensic Tool

The analysis of nuclear DNA sequences requires a considerable amount of DNA, but often only very small amounts are recovered at a crime scene. An extremely useful technique is the **polymerase chain reaction** (PCR) that allows an investigator to produce multiple copies of a particular preparation of DNA. This process is called **DNA amplification**. For example, PCR can produce sufficient human DNA from a small crime scene sample to test against a suspect's DNA to see if they match.

Although PCR focuses on comparisons of repetitive DNA, the entire genomes of many vertebrate species including humans as well as some invertebrates (e.g., annelids, mollusks, several insects) have been completely sequenced and many more are in progress. Furthermore, sequences of some genes have been used to construct phylogenetic trees to show the evolutionary relationships among species where complete genomes are not known. The use of multiple STRs has proven to be a reliable method for comparing humans

BOX 3.1 THE INNOCENCE PROJECT

Perhaps, the most dramatic demonstration of how DNA evidence can benefit justice comes from the efforts of national and several state organizations that have led to reversed convictions when DNA evidence was applied to a previous verdict. This has happened to people on "death row" and to people imprisoned unjustly for many years, even decades. How does this happen? When DNA evidence is not available or not used, circumstantial evidence, especially eyewitness accounts, often yield false convictions. The NAS Report (see Chapter 1) was especially critical of eyewitness evidence because several people viewing the same event often come up with disparate accounts of what they observed (Technical Working Group for Eye Witness Evidence, 1999; Wells and Loftus, 2003). Misidentifications via bite mark data and fingerprints as well as other forensic science errors (McMurtrie, 2005, 2010; Reardon, 2014) also have lead to many wrongful convictions (see Figure 3.10).

The Innocence Project (www.innocenceproject.org) centered with the Benjamin N. Cardozo School of Law, Yeshiva University in New York City, has at the time of this writing been responsible for reversal of over 325 convictions. The average time spent behind bars was 13 years. Several of those cleared were on death rows, and 70% were non-White people; almost all were male.

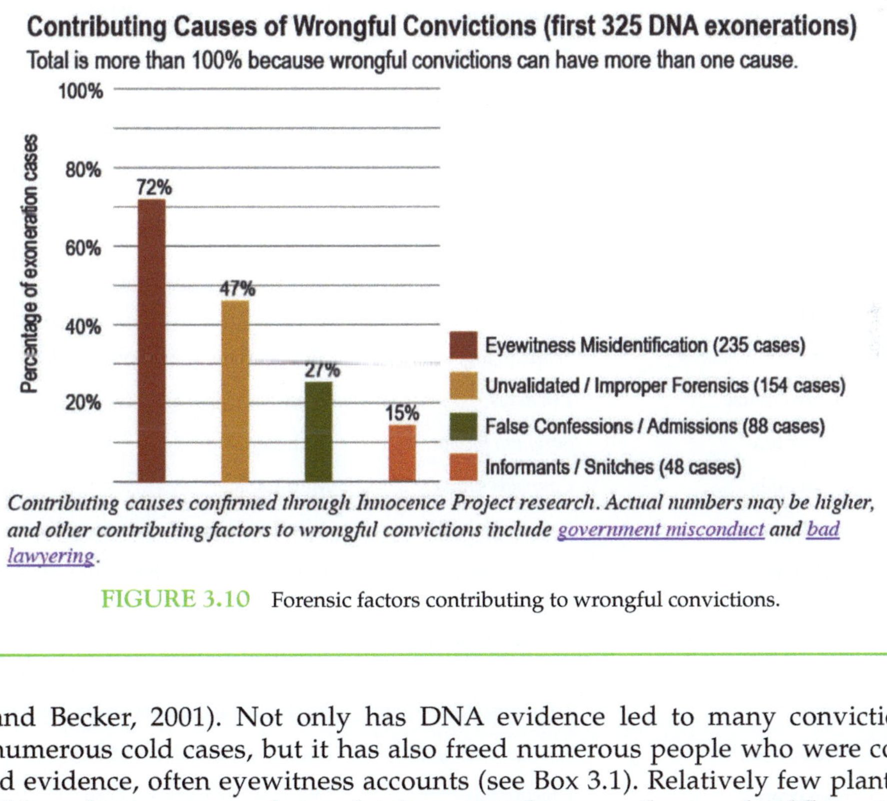

FIGURE 3.10 Forensic factors contributing to wrongful convictions.

(Butler and Becker, 2001). Not only has DNA evidence led to many convictions and cleared numerous cold cases, but it has also freed numerous people who were convicted on flawed evidence, often eyewitness accounts (see Box 3.1). Relatively few plant species (Table 3.1) have been sequenced completely among the many thousands of flowering plant species known. These are mainly common food plant species of considerable economic

TABLE 3.1 Some Plants with Known Nuclear Genomes

	Common name	Scientific name	Use
Food or food-related plants	Apple	*Malus × domesticus*	Fruit
	Pear	*Pyrus bretschneideri*	Fruit
	Chickpea (garbanzo bean)	*Cicer arietinum*	Legume
	Peach	*Prunus persica*	Fruit
	Common bean	*Phaseolus vulgaris*	Legume
	Sweet orange	*Citrus clementina*	Citrus fruit
	Papaya	*Carica papaya*	
	Tomato	*Solanum lycopersicum*	Vegetable (fruit)
	Potato	*Solanum tuberosum*	Vegetable (root)
	Grape	*Vitis vinifera*	Var. Pinot noir
	Pigeon pea	*Cajanus cajan*	Legume
	Cassava	*Manihot esculenta*	Source of tapioca
	Cucumber	*Cucumis sativus*	Inbred Chinese long 9930
	Melon	*Cucumis melo*	Cantaloupe, honeydew, etc.
	Watermelon	*Citrullus lanatus*	
	Woodland strawberry	*Fragaria vesca*	Related to garden varieties
	Date palm	*Phoenix dactylifera*	
	Banana (diploid form)	*Musa acuminata*	Store variety is triploid
	Rice	*Oryza sativa*	Grain
	Barley	*Hordeum vulgare*	Grain
	Einkorn wheat	*Triticum urartu*	"A" genome of hexaploid bread wheat
	Corn/maize	*Zea mays*	Grain
	Sorghum	*Sorghum bicolor*	Molasses source
Agricultural plants not used directly for food	Cannabis	*Cannabis sativa*	Marijuana
	Sugar beet	*Beta vulgaris*	Agricultural crop for sugar
	Rapeseed	*Brassica napa*	Agricultural crop for canola oil; var. *napobrassica* is rutabaga
	Other *Brassica* species	*Brassica oleracea*	In progress; other *Brassica* plants: brocolli, cauliflower, kale, Brussels sprouts, cabbage
	Flax	*Linum usitatissimum*	Agricultural crop for seeds, oil
	Castor bean	*Ricinus communis*	Source of castor oil and ricin
	Soybean	*Glycine max*	Legume; agricultural crop

TABLE 3.1 Some Plants with Known Nuclear Genomes—cont'd

	Common name	Scientific name	Use
	Cotton	*Gossypium raimondii*	"D" genome of *Gossypium hirsutum*, major crop
	Chocolate	*Theobroma cacao*	Varieties *croillo* and *matina*
	Foxtail millet	*Setaria italica*	C4 grass related to corn and sorghum
Wild plants	Barrel medic or barrel clover	*Medicago truncatula*	Legume used in genomic research
	Thale cress/mouse-eared cress	*Arabidopsis thaliana*	Brassicaceae; annual weed
	Lyre-leaved rock cress	*Arabidopsis lyrata*	Brassicaceae; annual weed
	Field mustard	*Brassica rapa*	*B. rapa* is garden turnip
	Moso bamboo	*Phyllostachys heterocycla*	Represents 70% of bamboo forests
	Shepherd's purse	*Capsella rubella*	2n; related to *Arabidopsis*
	Salt cress	*Thellungiella parvula*	Related to *Arabidopsis*
	No common name	*Amborella trichopoda*	Rare shrub of New Caledonia
	Columbine	*Aquilegia formosa*	Flower
	No common name	*Lotus japonicus*	Wild legume reference genome
Trees	Neem	*Azadirachta indica*	Mahagony tree relative
	Rose gum tree	*Eucalyptus grandis*	Tree
	Poplar	*Populus trichocarpa*	Tree
	Loblolly pine	*Pinus taeda*	Conifer tree (gymnosperm)
Primitive plants	Green alga	*Chlamydomonas reinhardtii*	Model organism for studies of photosynthesis, locomotion, etc.
	Moss	*Physcomitrella patens*	Model organisms for studies of plant evolution
	Lycophyte	*Selaginella moellendorffii*	Model organism for studies of comparative genomics

importance. Some of these species have very large genomes making the task a very arduous one. For example, the loblolly pine, a gymnosperm, has a genome that is roughly seven times greater than that of humans (Wegrzyn et al., 2014). Nevertheless, some progress has been made.

4.4 DNA Barcoding

A new technique called **DNA barcoding** is proving to be a useful technique for identifying plants (Sucher et al., 2012). DNA barcoding involves the use of a single gene to identify a given species through the comparison of nucleotide sequences in the DNA to that of the same gene in other species. This is in marked contrast to the multiple STR approach used for identifying individuals within a species. That is, it is simply used to identify a species taxonomically but not individuals. This approach has proven very effective in animals using the cytochrome oxidase 1 (*CO1*) gene. However, this gene in plants lacks the desired variability found in animals, and the chloroplast genes *rbcL* (ribulose 1,5, bisphosphate carboxylase) and *matk* (an open reading frame within the group II intron *trnK*) are commonly used instead (CBOL Plant Working Group, 2009). The matK protein is a degenerate form of a reverse transcriptase enzyme and is called a maturase. Although *rbcL* and *matK* have been used for phylogenetic analyses, their forensic application is simply to identify a species. However, sometimes it is only possible to identify the genus or perhaps only the family to which the plant belongs. Best discrimination was obtained for a variety of land plants (445 angiosperms, 38 gymnosperms, 67 crpytogams) by using both of these genes and some investigators recommend adding a third gene (*trnH-psbA*).

When attempting to identify sterile snakeroot roots as contraband, it was found that a smaller gene, *rpS16* intron, was more useful for DNA barcoding than the larger recommended genes due to the degraded nature of the DNA (Eurlings et al., 2013). Snakeroot, *Rauvolfia serpentina*, is an endangered species due to its overexploitation as a potent source of reserpine, an antipsychotic and antihypertensive drug. It is important to be able to differentiate the isolated roots of illegally collected snakeroot from related species in the same genus that are legal for harvesting. The *rp*S16 intron has proven useful in other studies as well (Wang et al., 2011).

4.5 What Is the Future for Uses of Plant DNA

Some success of DNA identification has been obtained for plants without known genomes using **amplified fragment length polymorphism markers (AFLP)** techniques. Bless et al. (2006) used AFLP to characterize red maple (*Acer rubrum*) with a successful identification of about 94% of the samples tested.

Australian workers have reported the use of chloroplast genes and mitochondrial genes in an alternative method to barcoding for identification of 100 species of Australian grasses (Ward et al., 2009) through progressive steps. However, the system is not simple to do like the human STR procedures. Use of microsatellite DNA among oak trees (*Quercus geminata*) allowed for identification of dried and fresh leaf material and connecting the material to individual trees (Craft et al., 2007). New techniques are allowing for extraction of chloroplast, mitochondrial, and repeated nuclear sequences from wood samples that also may have considerable promise for individualization (Deguilloux et al., 2002). Researchers in Poland developed a technique for species identification of mushrooms using **nuclear ribosomal DNA (nrDNA)** that yielded a much higher percentage of positive identifications than the conventional method of using spores to identify the species (Kowalezyk et al., 2015).

Methods have been developed for distinguishing among various species of *Cannabis* (Howard et al., 2008; Johnson et al., 2013; Valverde et al., 2014). Some studies can distinguish among varieties of a single species of *Cannabis*, which especially will be useful to forensic scientists (Mendoza et al., 2009).

Identification of most plant species still relies heavily on anatomy and morphology when significant portions of the plant are available. Hopefully, DNA techniques will become routine enough and inexpensive so that they can be readily applied in the future to tiny fragments containing chloroplast DNA, mitochondrial DNA, or nuclear DNA.

References

Baker, H.G., Stebbins, G.L., 1965. The Genetics of Colonizing Species. Academic Press, New York.

Barr, C.M., Neiman, M., Taylor, D.R., 2005. Inheritance and recombination of mitochondrial genomes in plants, fungi and animals. The New Phytologist 168, 39–50.

Billings, W.D., 1985. The historical development of physiological plant ecology. In: Chabot, B. (Ed.), Physiological Ecology of North American Plant Communities. Chapman and Hall, New York, pp. 1–15.

Birsky Jr., C.W., 1995. Uniparental inheritance of mitochondrial and chloroplast genes: mechanisms and evolution. Proceedings of the National Academy of Sciences of the United States of America 92, 11331–11338.

Bless, C., Palmeter, H., Wallace, M.M., 2006. Identification of *Acer rubrum* using amplified fragment length polymorphism. Journal of Forensic Science 51, 31–38.

Blunt, W., 1971. Linnaeus: The Complete Naturalist. William Collins Sons and Co, London.

Bock, J.H., 1973. The salient features of the International Code of Botanical Nomenclature. The Biologist 55, 13–23.

Butler, J.M., Becker, C.H., 2001. Improved Analysis of DNA Short Tandem Repeats with Time-of-flight Mass Spectrometry. Science and Technology Research Report, NCJ 188292. U.S. Department of Justice. 76pp.

CBOL Plant Working Group, 2009. A DNA barcode for land plants. Proceedings of the National Academy of Science of the United States of America 106, 12794–12797.

Clay, R.S., Court, T.H., 1932. The History of the Microscope: Compiled from Original Instruments and Documents up to the Introduction of the Achromatic Microscope. Griffin, London.

Clements, F.E., 1913. The alpine laboratory. Science 3, 327–328.

Craft, K.J., Owens, J.D., Ashley, M.V., 2007. Application of plant DNA markers in forensic botany: genetic comparisons of Quercus evidence leaves to crime scene trees using microsatellites. Forensic Science International 165, 64–70.

Deguilloux, M.-F., Pemonge, M.-H., Petit, R.J., 2002. Novel perspectives in wood certification and forensics: dry wood as a source of DNA. Proceedings of the Royal Society of London B 269, 1039–1046.

Esau, K., 1977. Anatomy of Seed Plants. John Wiley & Sons, New York.

Eurlings, M.C.M., Lens, F., Pakusza, C., Peelen, T., Wieringa, J.J., Gravendeel, B., 2013. Forensic identification of Indian snakeroot (*Rauvolfia serpentina* Benth. ex Kurz) using DNA barcoding. Journal of Forensic Sciences 58, 822–830.

Evert, R.F., Eichhorn, S.E., 2006. Esau's Plant Anatomy: Meristems, Cells, and Tissues of the Plant Body: Their Structure, Function, and Development. John Wiley & Sons, Hoboken, NJ.

Gleason, H.A., 1926. The individualistic concept of the plant association. Torrey Botanical Club Bulletin 53, 7–26.

Goodland, R., 1975. History of "Ecology". Science 188, 313.

Grew, N., 1672. The Anatomy of Vegetables with a General Account of Vegetation Founded Thereon. Spenser Hickman, London.

Hooke, R., 1665. Micrographia. Johannis Martyn, London.

Howard, C., Gilmore, S., Robertson, J., Peakall, R., 2008. Developmental validation of a *Cannabis sativa* STR multiplex system for forensic analysis. Journal of Forensic Sciences 53, 1061–1967.

Jenny, H., 2011. Factors of Soil Formation: A System of Quantitative Pedology. Dover Press, Mineola, New York.

Johnson, C.E., Premasuthan, A., Satkoski Trask, J., Kanthaswamy, S., 2013. Species identification of *Cannabis sativa* using real-time quantitative PCR (qPCR). Journal of Forensic Sciences 58, 486–490.

Kowalezyk, M., Sekula, A., Mleczko, P., Olszowy, Z., Kujawa, A., Zubek, S., Kupiec, T., 2015. Practical aspects of genetic identification of hallucinogenic and other poisonous mushrooms for clinical and forensic purposes. Croatian Medical Journal 56, 32–40.

Ledbetter, N., Porter, K., 1964. Science 144, 872.

Linnaeus, C., 1753. Species Plantarum. Laurentius Salvius, Stockholm.

Malpighi, M., 1686. Philosophi and Medici Bononiensis e Regia Societate Opera Omni Incisis Illustrata: Tomis Duobus Comprehensa, Quorum Catalogum Sequens Pagina Exhibit. Prostant apud Robertum Scott, London.

McNeill, J., Barrie, F.R., Buck, W.R., Demoulin, V., Greuter, W., Hawksworth, D.L., Herendeen, P.S., Knapp, S., Marhold, K., Prado, J., Prud'Homme Van Reine, W.F., Smith, G.F., Wiersema, J.H., Turland, N.J., Secretary of the Editorial Committee, 2012. International Code of Nomenclature for Algae, Fungi and Plants (Melbourne Code) Adopted by the Eighteenth International Botanical Congress Melbourne, Australia. Koeltz Scientific Books, Koenigstein, Germany.

Mendoza, M.A., Mills, D.K., Lata, H., Chandra, S., ElSohly, M.A., Almirall, J.R., 2009. Genetic individualization of *Cannabis sativa* by a short tandem repeat multiplex system. Analytic and Bioanalytical Chemistry 393, 719–726.

Moran, J., 1973. Printing Presses: History and Development from the Fifteenth Century to Modern Times. University of California Press, Berkeley.

McMurtrie, J., 2005. The role of the social sciences in preventing wrongful convictions. American Criminal Law Review 42, 1271–1287.

McMurtrie, J., 2010. Swirls and whirls: litigating post-conviction claims of fingerprint misidentification after the NAS report. Utah Law Review 2010, 267–297.

Neale, D.B., Marshall, K.A., Sederoff, R.R., 1989. Chloroplast and mitochondrial DNA are paternally inherited in *Sequoia sempervirens* D. Don Endl. Proceedings of the National Academy of Sciences of the United States of America 86, 9347–9349.

Neale, D.B., Sederoff, R.R., 1989. Paternal inheritance of chloroplast DNA and maternal inheritance of mitochondrial DNA in loblolly pine. Theoretical and Applied Genetics 77, 212–216.

Norris, D.O., Bock, J.H., 2000. Use of fecal material to associate a suspect with a crime scene: report of two cases. Journal of Forensic Science 45, 178–181.

Pavord, A., 2005. The Naming of Plants: The Search for Order in the World of Plants. Bloomsbury Publishing, New York.

Pound, R., Clements, F.E., 1900. The Phytogeography of Nebraska. University of Nebraska Press, Lincoln.

Ruska, E., 1986. The emergence of the electron-microscope: connection between realization and 1st patent application, documents of an invention. Journal of Ultrastructure and Molecular Structure Research 95, 3–28.

Reardon, S., 2014. Faulty forensic science under fire. Nature 506, 13–14.

Stebbins Jr., G.L., 1950. Variation and Evolution in Plants. Columbia University Press, New York.

Sucher, N.J., Hennell, J.R., Carles, M.C., 2012. DNA fingerprinting, DNA barcoding, and next generation sequencing technology in plants. Methods in Molecular Biology 862, 13–22 Springer (Humana Press), New York.

Tansley, A.G., 1947. The early history of plant ecology in Britain. Journal of Ecology 35, 130–137.

Technical Working Group for Eye Witness Evidence, 1999. Eye Witness Evidence. A Guide for Law Enforcement. US Department of Justice. pp. 37–48.

Valverde, L., Lischka, C., Scheiper, S., Nedele, J., Challis, R., de Pancorbo, M.M., Pfeiffer, H., Köhnemann, S., 2014. Characterization of 15 STR cannabis loci: nomenclature proposal and SNPSTR haplotypes. Forensic Science International: Genetics 9, 61–65.

Ward, J., Gilmore, S.R., Robertson, J., Peakall, R., 2009. A grass molecular identification system for forensic botany: a critical evaluation of the strengths and limitations. Journal of Forensic Sciences 54, 1254–1260.

Wang, D.-Y., Zhang, Q., Liu, Y., Zhi-Fu, L., Shao-Xiang, Z., Meng-Xiang, S., Sodmergen, 2010. The levels of male gamete mitochondrial DNA are highly regulated in angiosperms with regard to mitochondrial inheritance. The Plant Cell 22, 2402–2416.

Wang, Y., Wang, W., Tong, W., Zhao, W., 2011. Analysis of chloroplast ribosomal subunit S16 (rpS16) intron sequences in Morus (Urticales: Moraceae). African Journal of Biotechnology 10, 17695–17699.

Wegrzyn, J.L., Liechty, J.D., Stevens, K.A., Wu, L.-S., Loopstra, C.A., et al., 2014. Unique features of the loblolly pine (*Pinus taeda* L.) megagenome revealed through sequence annotation. Genetics 196, 891–909.

Wells, G.L., Loftus, E.F., 2003. Eyewitness memory for people and events. In: Goldstein, A.M., Weiner, I.B. (Eds.), Handbook of Physiology, vol. 11. John Wiley & Sons, New York, pp. 149–160. http://dx.doi.org/10.1002/0471264385.wei1109. Published Online: April 15, 2003.

CHAPTER

4

Forensic Plant Anatomy

In this chapter, we provide a background for understanding plant anatomy as well as a brief overview of food processing by humans. Use of gastrointestinal (GI) contents in determining time of death or connecting a suspect to a crime scene is discussed. Additionally, we provide information for collection and processing of samples for forensic analyses.

1. SOME PLANT BASICS

The following account is a brief orientation to the basic features of plants that can be utilized in forensic analyses with emphasis on the flowering plants that we commonly consume as food. The reader should consult some of the many plant science textbooks, such as those listed in Chapter 3, for detailed accounts of all plant groups.

1.1 Types of Plants

We can separate the plant world into two major categories: seed plants (flowering plants and conifers) and nonseed plants. The latter group consists of more primitive plants such as algae, bryophytes, and ferns. Some examples of human foods are found here, but they do not form a major part of the diet in most cases. Some algae are eaten directly and algal extracts are widely used as thickeners and emulsifiers in many food products. Among the fungi, mushrooms are eaten and have been responsible for accidental and criminal poisonings (see Chapter 1). Young ferns (fiddleheads) are a delicacy in only certain regions of North America but are uncommon in the diets of most North Americans.

The seed plants consist of the **gymnosperms** (including conifers: pines, spruces, firs, etc.). The term "gymnosperm" means "*gymnos*, naked; *sperm*, seed," emphasizing that their seeds are produced on the open surfaces of cones or other structures. The conifers produce seeds on cones, whereas the flowering plants or **angiosperms** (*angio-*, covered; seeds are found within a fruit) form theirs within the ovaries of flowers. The ovaries of flowers develop into fruits. It is in the angiosperms that most of our food plants are found.

Flowering plants were originally subdivided into two groups based on their development, anatomy and morphology. A mature seed contains a tiny embryo that consists of a primitive rootlike structure (**radicle**), a primitive stem, and one or two primitive leaves called **cotyledons** (Figure 4.1). After the germinating, seed reaches the surface, the cotyledons or "seed leaves"

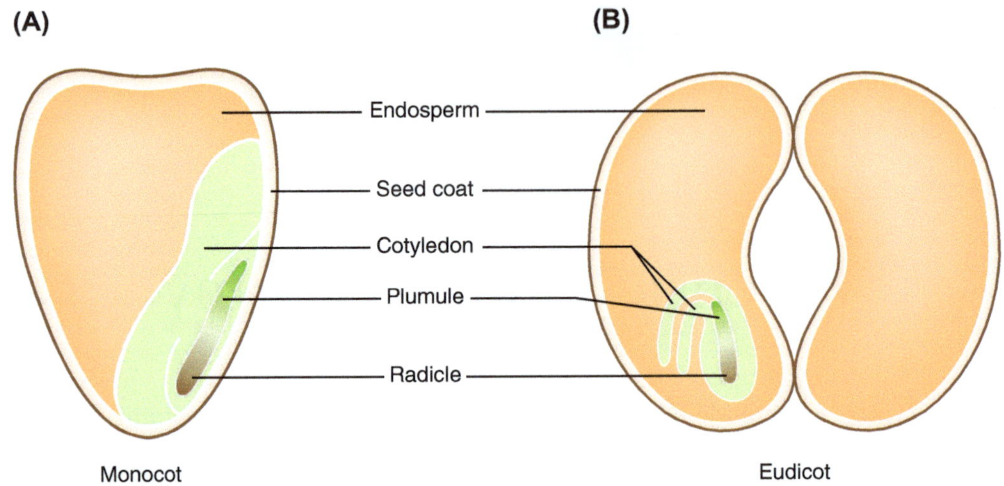

FIGURE 4.1 Monocot (A) and eudicot (B) plant embryos. (1) Cotyledon(s), (2) endosperm (stored food), (3) plumule that gives rise to the shoot, (4) radicle that gives rise to the root system, (5) seed coat.

will begin photosynthesis for initial growth until true leaves appear. Plants with a single cotyledon are termed **monocots** (e.g., grasses, sedges) and represent about 22% of all flowering plants. Plants with two cotyledons originally were termed **dicots**. Most of the plants with two cotyledons now are termed **eudicots** (see Simpson, 2010), and they represent about 75% of all flowering plants (e.g., many flowers, most edible fruits and vegetables, and all flowering shrubs and trees). The remaining original dicots (3% of flowering plants) have been reclassified into several small groups, but none are important food plants (Figure 4.2). There are numerous structural features that separate the monocots from the eudicots (see Table 4.1).

Certain plant families deserve special mention because of their widespread significance to human diets universally. Among the monocots, the grass family, Poaceae, furnishes the staffs of life for most of the world's cultures in the form of three species: *Oryza sativa*, common name rice; *Zea mays*, common name corn or maize; and *Triticum*[1] spp., common name wheat or corn.[2] The common names of two of these three grains illustrate the weakness of using a common name to designate a plant rather than its scientific name.

A second highly significant family is the eudicot legume family, Fabaceae. This family provides vegetable protein in peas and beans to most cultures.

Another eudicot family of special significance especially in the northern hemisphere is the rose family, Rosaceae. The woody members of this family provide us with many important fruits including cherries, apples, peaches, pears, and plums.

A fourth family, the eudicot mustard family, Brassicaceae, illustrates a striking result of plant genetics and artificial selection that has happened for other groups as well. One species in this family *Brassica oleracea*, is made up of several vegetables that are valued for their leaves

[1] Although Linnaeus named only five species of wheat, today more than 20 species of "wheat" are recognized by taxonomists.

[2] The common name "corn" was used in England for what North Americans call "wheat," probably stemming from the German use of the word *kern* for seed. American "corn" was simply known as "maize" in England until recently.

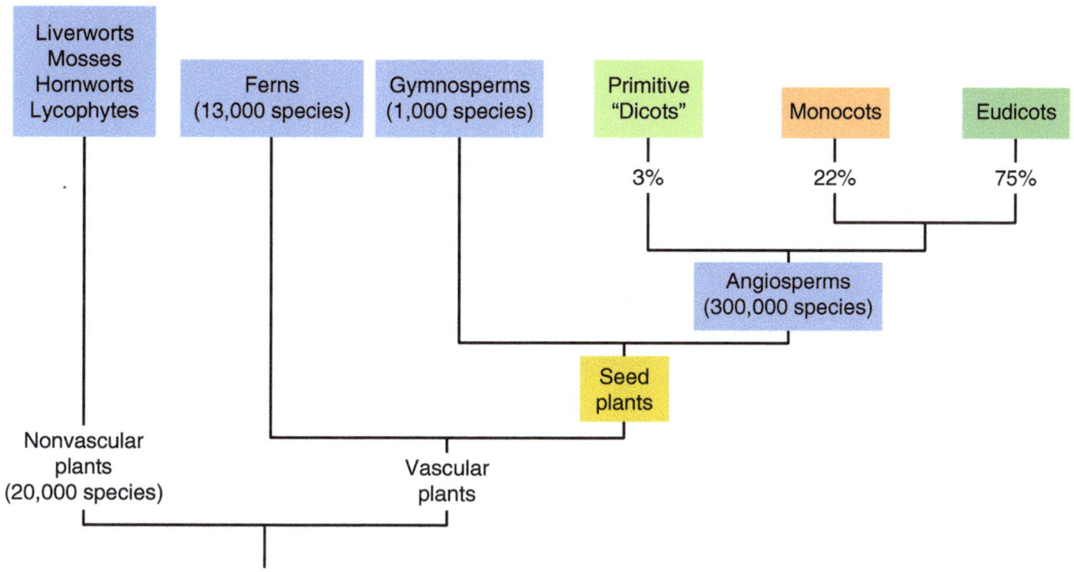

FIGURE 4.2 Simplified diagram of the evolution of terrestrial plants (fungi and algae not included). Most forensic plant science is focused on the eudicots (approximately 75% of 300,000 described extant angiosperm species) and conifers (most of the 1000 species of gymnosperms).

TABLE 4.1 Comparison of Some Features of Monocot and Eudicot Flowering Plants

Monocots	**Eudicots**
Seeds having one cotyledon	Seeds having two cotyledons
Pollen grain has one furrow or aperture	Pollen grain has three furrows or apertures
Leaf venation mostly parallel	Leaf venation mostly netlike or reticulate
Leaf epidermis with rectangular cells	Leaf epidermis with angular or jigsaw puzzle-shaped cells
Leaf usually wraps around stem	Leaf usually attached to stem by a short stalk (petiole)
Fibrous root system (adventitious roots)	Typically a central taproot system
Flower parts usually in threes or multiples of three	Flower parts usually in fours or fives or in multiples of four or five
Stems usually herbaceous	Stems woody or herbaceous
22% of all angiosperms	75% of all angiosperms

and/or flowers. Forage kale is thought to be the ancestral type for this group. *Brassica oleracea* is also the correct Latin species name for kale, cabbage, cauliflower, broccoli, and kohlrabi (Table 4.2). They evolved with relatively few genetic mutations from a common ancestor, forage kale, but they overall are so genetically indistinct that they are separated only by varietal names. Around 1750, the Brussels sprout mutation of *B. oleracea* showed up in a garden in Belgium (Simpson and Ogorzaly, 1995). Obviously, identification of these plants through genetic analyses would be challenging. They are also difficult to separate using microscopic observation.

TABLE 4.2 Some Common Vegetables of the Species *Brassica oleracea* and *Cucurbita pepo*

Common name	Scientific name
Forage kale (ancestral)	*Brassica oleracea*
Broccoli	*Brassica oleracea* var. *botrytis*
Brussel sprouts	*Brassica oleracea* var. *gemmifera*
Cabbage	*Brassica oleracea* var. *capitata*
Cauliflower	*Brassica oleracea* var. *cauliflora*
Kale	*Brassica oleracea* var. *acephala*
Kohlrabi	*Brassica oleracea* var. *gongyloides*
Acorn squash	*Cucurbita pepo* var. *turbinate*
Yellow crookneck	*Cucurbita pepo* var. *torticollia*
Pumpkin	*Cucurbita pepo* var. *pepo*
Scallop	*Cucurbita pepo* var. *clypeata*
Yellow summer squash	*Cucurbita pepo* var. *cylindrica*
Spaghetti squash	*Cucurbita pepo* var. *fastigatus*

In order to identify angiosperm food plants in our diets after consumption, it is essential to have some knowledge of plant cell types. The cellular composition of flowering plants pertinent to the study of food plants is described ahead.

1.2 Flowering Plant Cell Types

Flowering plants consist of organized tissues composed of one or more cell types that are organized into the three tissue systems described here. Each tissue system consists of specialized cells organized in unique patterns. They make up the leaves, flowers, stems, roots, fruits, and seeds that humans commonly consume as food. Hence knowledge of plant tissues can be of special importance for forensic analyses of GI contents and fecal matter.

1.2.1 Tissue System 1: The Dermal System

The **dermal system** includes epidermal cells and associated trichomes. **Epidermal cells** occur at the surface of leaves and stems and roots as well as flowers and fruits. In monocots, they are typically rectangular in shape, whereas in eudicots they resemble jigsaw puzzle pieces (Figure 4.3). Epidermal cells typically lack organelles called chloroplasts that contain the green photosynthetic pigment, chlorophyll. A pair of specialized **guard cells** (epidermal cells with chloroplasts) surrounds an opening, the **stoma** (*pl.*, stomata), forming a **stomate** (Figure 4.4). The guard cells determine the size of the stoma along the leaf surface thus controlling the exchange of air between the environment and the inner parts of the leaf where photosynthesis occurs.

Trichomes are epidermal specializations consisting of one or more cells that protrude from the epidermis (Figures 4.5 and 4.6). Some are simple, whereas others may be branched. They are found, for example, on the pods of green beans and okra as well as on peaches and

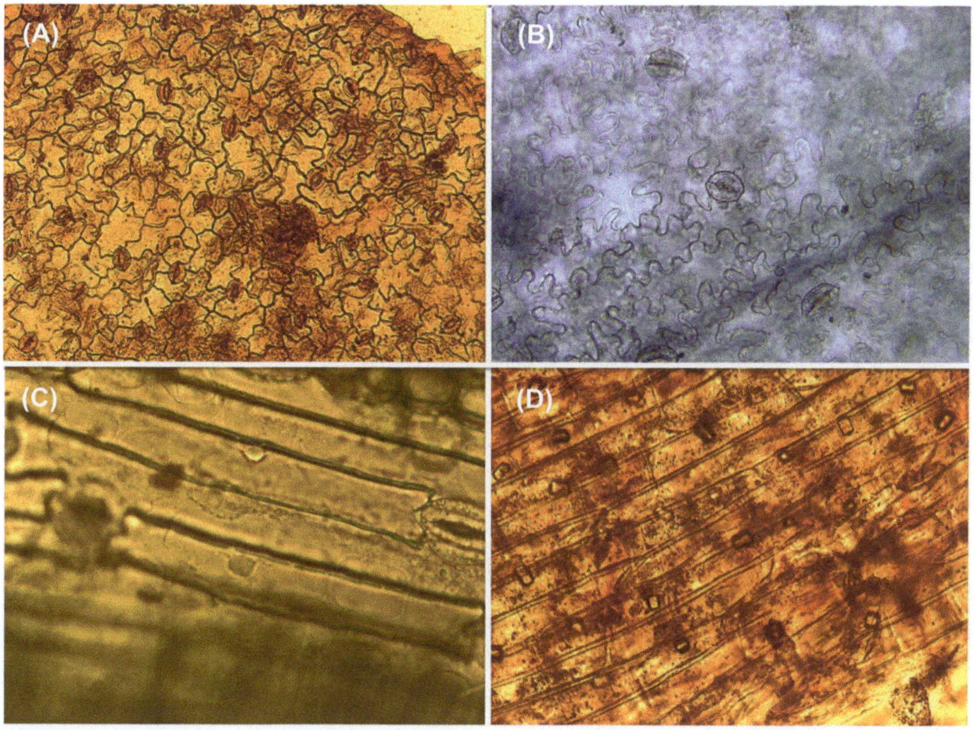

FIGURE 4.3 Epidermal cells of eudicots (A, B) and monocots (C, D). Note that the eudicot epidermis looks like a jigsaw puzzle, whereas the monocot epidermis resembles a brick wall. (A) Spinach epidermis stained with safranin O, 10×. (B) Lettuce epidermis unstained, 10×. (C) Chive epidermis, unstained, 40×. (D) Leek epidermis underlain with parenchymal cells, stained with safranin O, 25×. *Photomicrographs by author.*

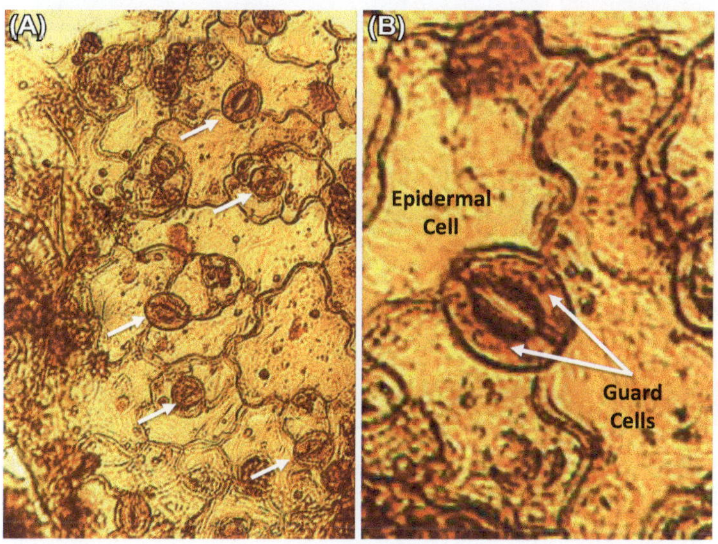

FIGURE 4.4 Stomate apparatus. (A) Leaf epidermis of arugula showing multiple stomata (arrows). Stained with safranin, 25×. (B) Note that guard cells stain darker than the epidermal cells in part because they contain chloroplasts that are absent in the surrounding epidermal cells. *Photomicrographs by author.*

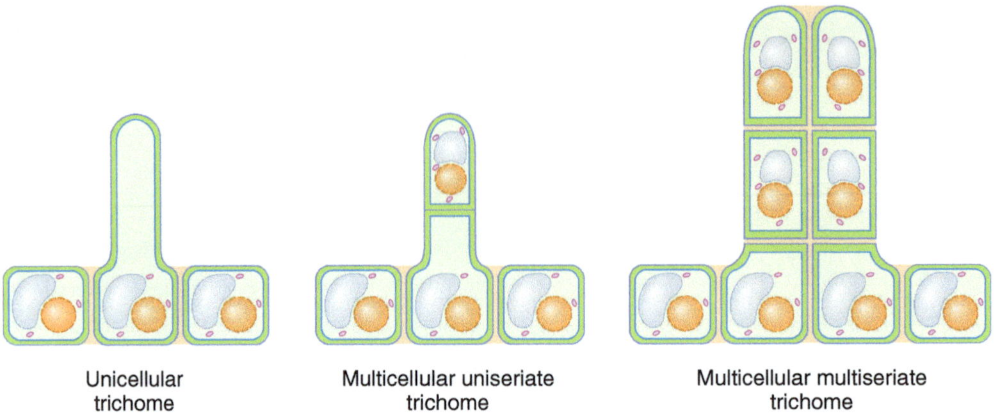

FIGURE 4.5 Types of trichomes.

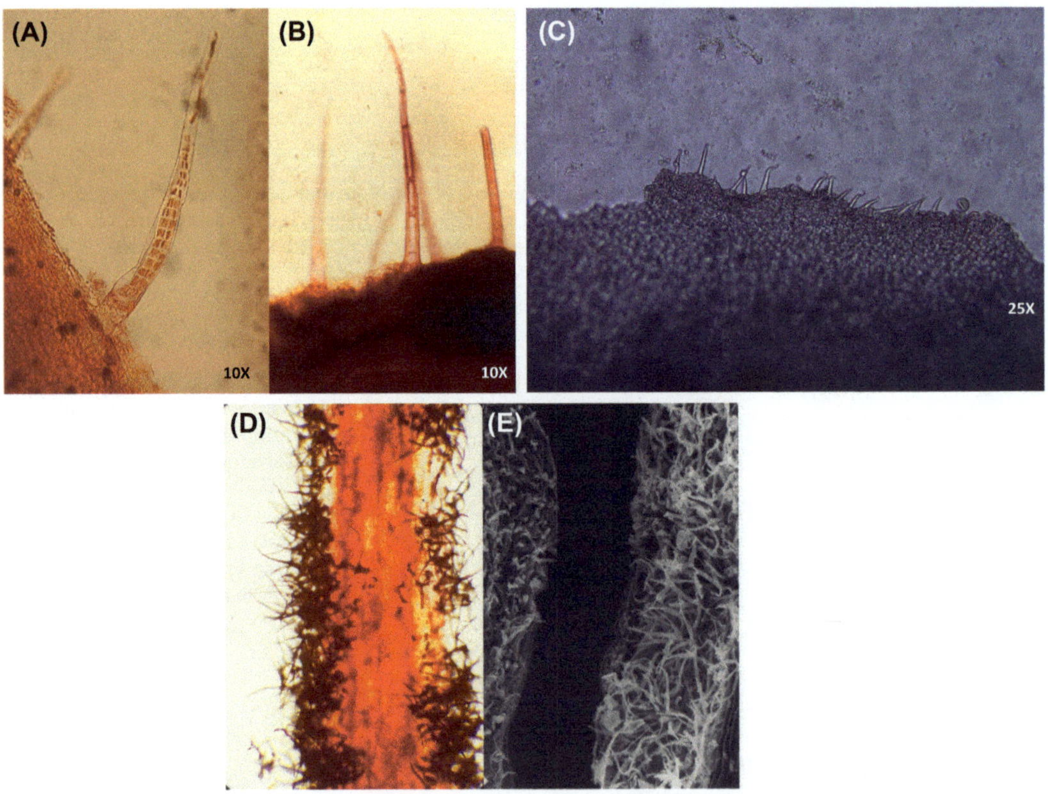

FIGURE 4.6 Trichomes. (A) Okra (multicellular, multiseriate); (B) Oregano (multicellular, uniseriate); (C) Fig (unicellular); (D) and (E) Rosemary (D&E, unicellular, triseriate). *Light microscope photomicrographs by author. SEM courtesy of Dr. Meredith A. Lane.*

kiwis and along the stalks (petioles) of rosemary leaves. Trichomes of some plants also may be secretory (i.e., glandular) such as those on *Cannabis*.

1.2.2 Tissue System 2: Parenchyma, Collenchyma, and Sclerenchyma[3]

Parenchymal cells (Figure 4.7) function in photosynthesis and/or as storage sites. Various nutrients may be stored within the intracellular plastids or vacuoles of parenchymal cells. They occur in leaves, roots, and stems, and make up the bulk of fleshy fruits. Parenchymal cells have relatively thin cell walls. Some parenchymal cells differentiate into collenchymal and sclerenchymal cells.

Collenchymal cells (Figure 4.8) are elongated cells and generally have high water content. They exhibit thickened cell walls where they contact neighboring cells providing for rigidity. Collenchymal cells give support to plants, especially in actively growing regions. For example, collenchymal cells are a major component of the fibers that characterize the stalks of celery.

Sclerenchymal cells (Figure 4.9) have very thick cell walls, usually strengthened by the addition of chemicals called **lignins**. They occur as heavy elongate **fibers** as in stems or as

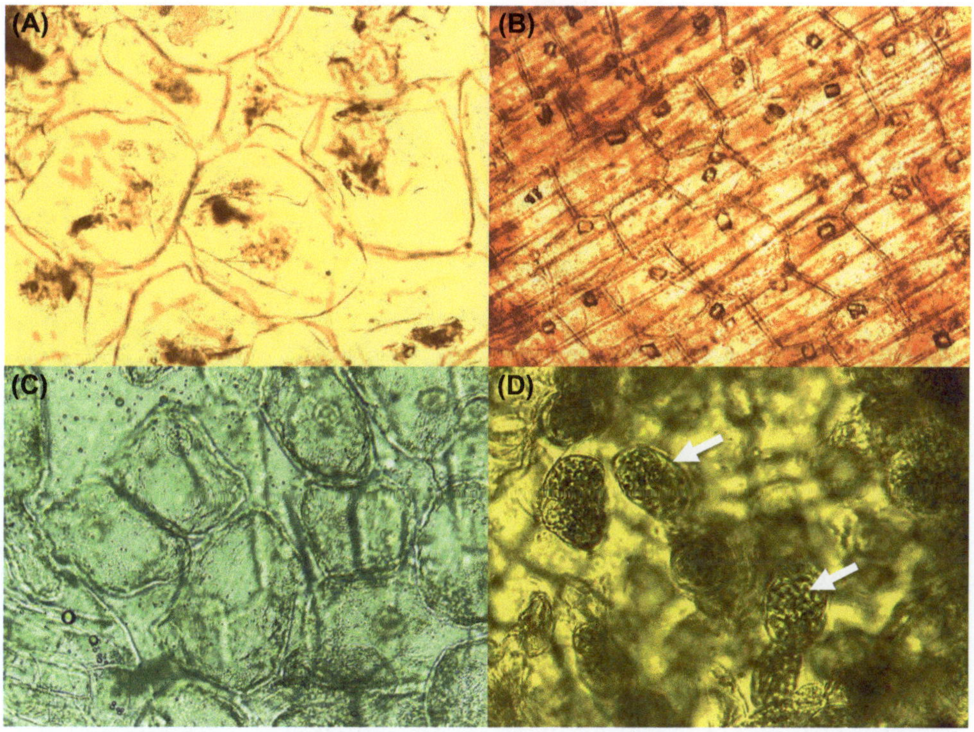

FIGURE 4.7 Parenchyma. (A) Tomato fruit storage parenchymal cells, 10×. (B) Leek parenchymal cells underlying rectangular epidermal cells, 10×. (C) Carrot root parenchyma, 25×. (D) Spinach photosynthetic parenchymal cells (arrows). *Photomicrographs by author.*

[3]These three cell types (parenchyma, collenchyma, and sclerenchyma) are sometimes referred to as "the ground system." The term appears to be a mistranslation of the German **grund** that can mean "fundamental" or "basic" as well as "ground."

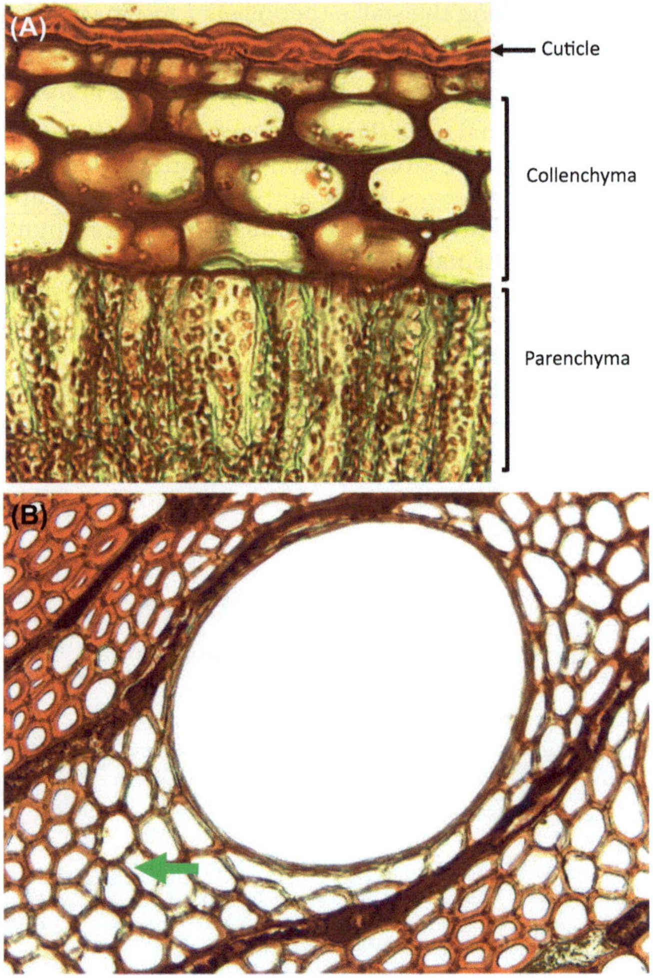

FIGURE 4.8 Collenchymal cells. (A) Present in a xerophylic leaf cross section beneath a single layer of epidermal cells covered with a waxy cuticle. The elongated cells at the bottom are parenchymal cells. (B) Present as component of stems (arrow). Note they surround a large vessel in this cross section of oak stem. *Photomicrographs by author.*

shorter **sclereids** (also called stone cells). Sclereids give the flesh of pears a slightly gritty texture. They are common in blueberries where along with tiny seeds they contribute to the texture of the fruit. Generally, fibers and sclereids provide support and protection for plants. We usually don't eat plant parts that have many sclerenchymal cells.

1.2.3 Tissue System 3: Vascular Tissue

Two kinds of elongated vascular cells characterize both gymnosperms and flowering plants. One kind, **xylem** (Figures 4.10 and 4.11), conducts water and dissolved minerals from the roots to the leaves and forms the bulk of the vascular tissue. Xylem consists only of **tracheids** in ferns, most gymnosperms, and the more primitive flowering plants. Most of the flowering plants have tracheids as well as large diameter conduction cells called **vessels** that also are important for water conduction. Elongated fibers (sclerenchyma) are also found with the conducting cells and provide additional support in the xylem.

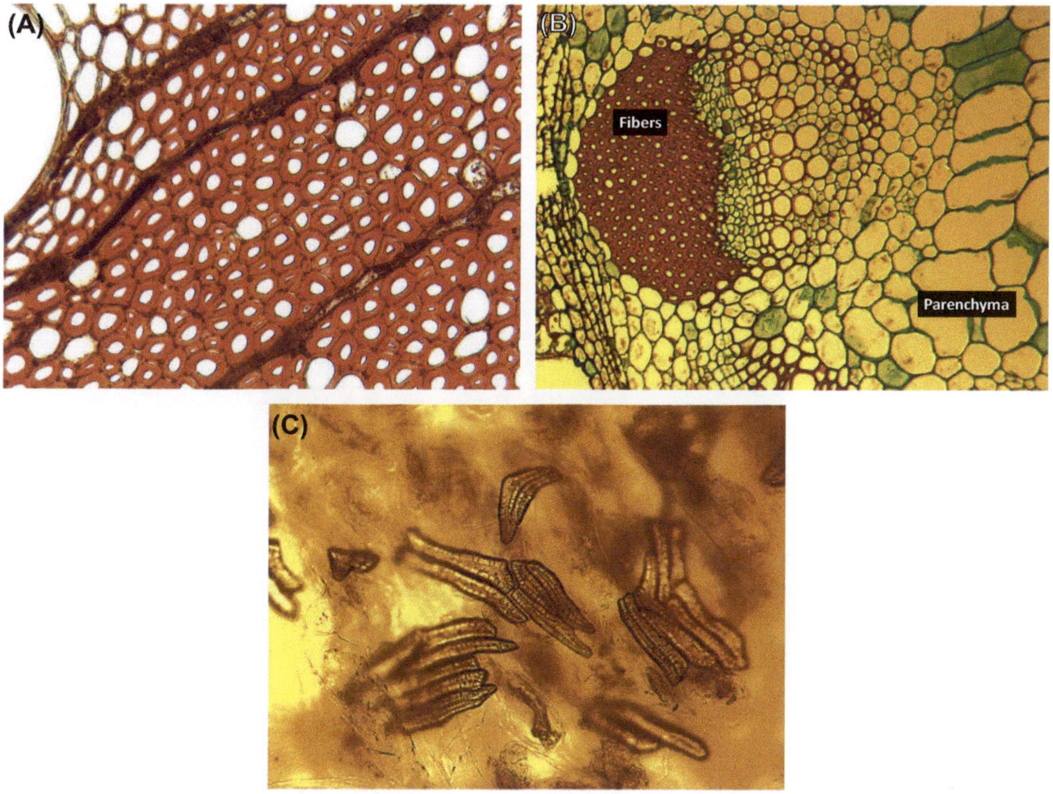

FIGURE 4.9 Sclerenchymal cells. (A) Present in a xerophylic leaf cross section, 40×. (B) Present as fibers in cross section of oak stem, 40×. Note they surround a large vessel. (C) Sclerenchymal cells (sclerids) present in fruit of blueberry. *Photomicrographs by author.*

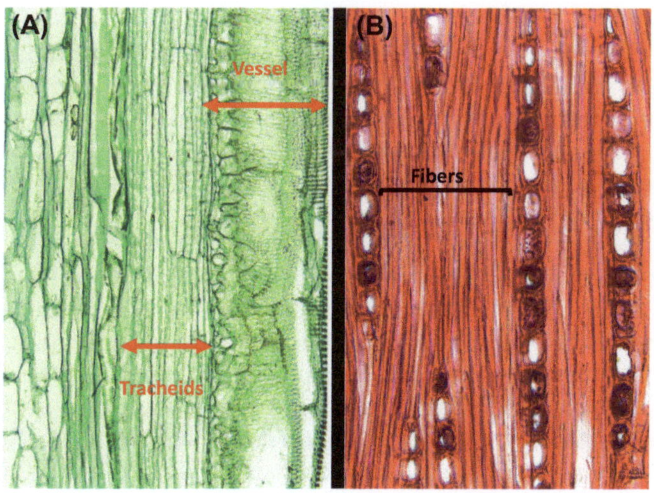

FIGURE 4.10 Xylem tissue. (A) Low power (10×) view of a vessel and tracheids in eudicot vascular tissue of squash (*Cucurbita*). (B) Low power view of pine stem, a conifer lacking vessels. *Photomicrographs by author.*

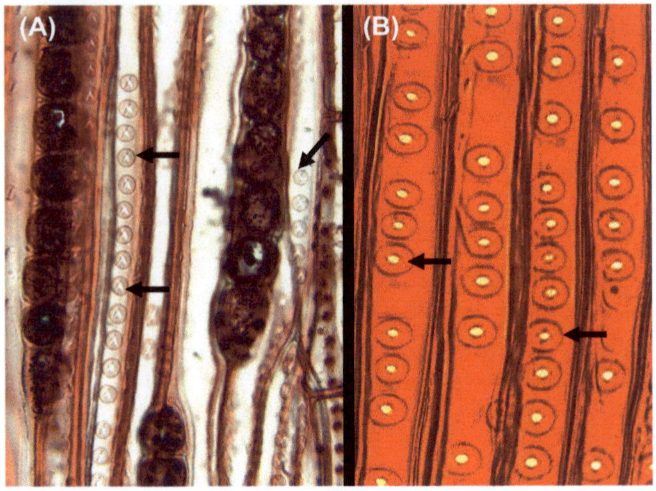

FIGURE 4.11 Tracheid pits (arrows) seen with high magnification (40×). (A) Oak. (B) Pine. *Photomicrographs by author.*

Functioning xylem cells are all dead and pits or perforations in the cell walls allow water to move from one hollow cell to the next forming continuous tubes from roots to leaves. In stems, the xylem collects to form bundles that continue into the leaves as veins. In monocots, the leaf veins appear parallel to the long axis of the leaf. Furthermore, the base of the monocot leaf wraps around the stem. The veins in eudicots appear as branching trees of vascular tissue, and the leaves are attached to the stem via a short stalk, the **petiole**.

The second kind of vascular tissue is **phloem**. Unlike xylem, phloem cells are living and function to move organic materials up and down the stem of a plant. Phloem cells actually make only a minor contribution to the foods we eat and are not important for identification purposes.

1.2.4 Inclusions

In addition to possessing a variety of different cell types, some cells may accumulate crystalline substances that provide another characterizing feature. Crystals may come in a variety of sizes and shapes and the pattern of their distribution and composition may vary greatly from species to species. Depending on their composition, they may occur in **prismatic shapes**, as **druse** or as needlelike **raphides** (Figures 4.12 and 4.13) or simply as irregular clumps. Clumps of raphides may be found in vacuoles within cells as in kiwi (Figure 4.14).

Additionally, many plants produce internal silicon dioxide inclusions called **phytoliths** (Hart, 2015). Phytoliths (Figure 4.15) can be used for identification because they conform to actual structures within the plant and hence are characteristic of different species (see Schneck, 2004; Piperno, 2006; Hart, 2015). When a plant dies, the phytoliths persist in the soil and also may be found in coprolites (fossil feces). Phytoliths have been used by archeologists and paleoecologists to reconstruct the diets of primitive peoples and extinct animals, respectively (Piperno, 2006). Phytoliths also have been used successfully to compare soils and connect a suspect's vehicle to a particular location (Schneck, 2004).

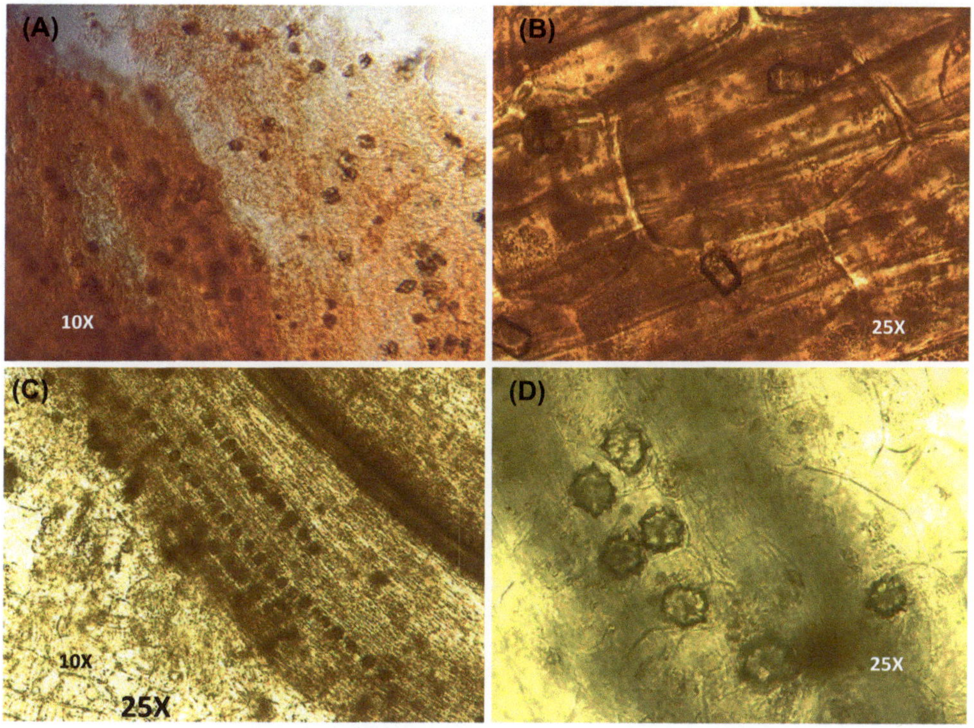

FIGURE 4.12 Crystals. (A) Druse crystals in fig. (B) Rectangular crystals in leek. (C and D) Okra druse crystals at two magnifications. *Photomicrographs by author.*

FIGURE 4.13 Raphide phytoliths from pineapple. *Photomicrograph by author.*

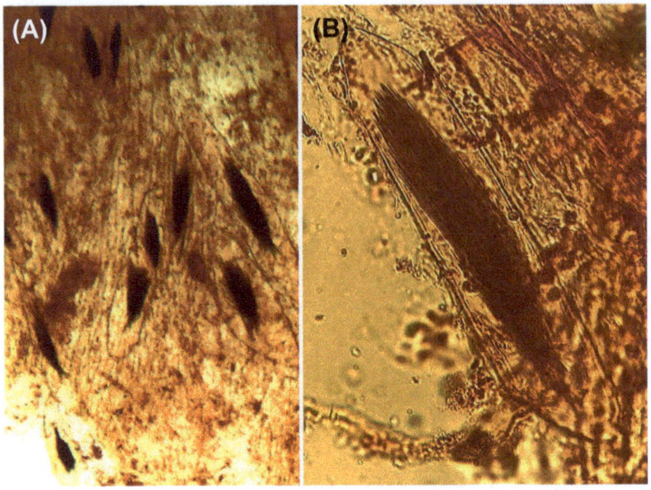

FIGURE 4.14 Inclusions in the fruit of kiwi consisting of bundles of raphides within a specialized cell. (A) 10×, (B) 25× of single cell. Note the individual raphides making up the compact inclusion. *Photomicrographs by author.*

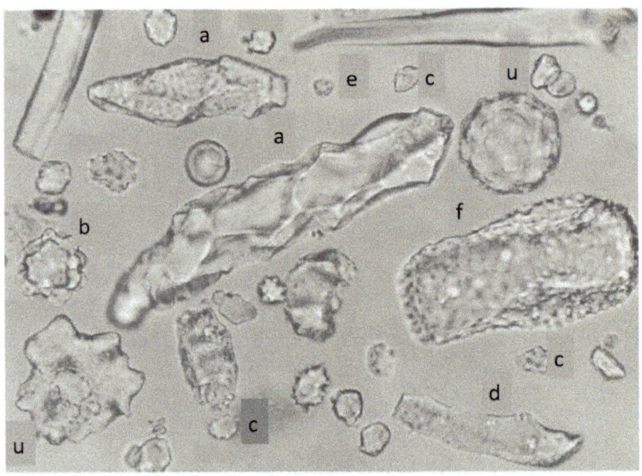

FIGURE 4.15 Phytoliths can be used to identify families, genera, and even species of plants. From a Panama soil sample: a, magnolia; b, member of the arrowroot family (Marantaceae); c, palms; d/e, described only by shape; f, trees; u, unknown. *Modified with permission from Hall (2015).*

Some plants may contain distinctive **starch grains** that are especially visible with polarized light (Figure 4.16). Starch is a polymer of glucose formed to produce amylose strands and/or amylopectin strands. The presence of starch can be detected by the addition of a potassium iodide solution (see Appendix IV) causing plastids containing starch high in amylose to turn a bluish black (Figure 4.17), whereas those high in amylopectin starch stain yellow to reddish brown.

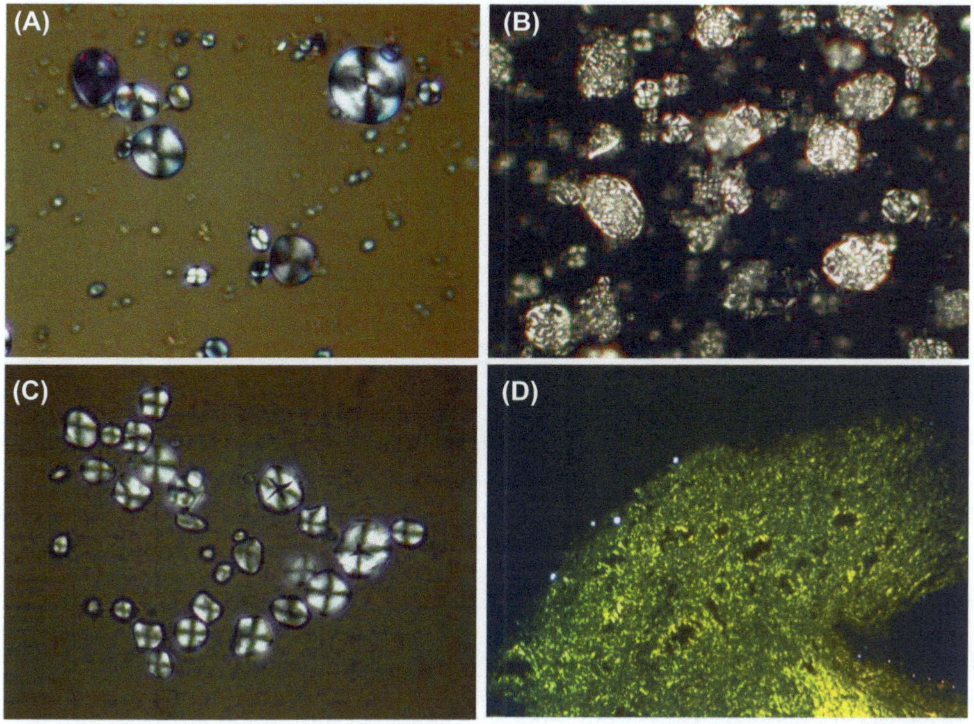

FIGURE 4.16 Starch grains appear as white spheres when viewed with polarized light. (A) Corn starch. (B) Oat starch. (C) Wheat starch. (D) Starch grains on the surface of human autopsy tissue using polarized light at low magnification. *Figures (A–C) reprinted with permission from W.M. Schneck (2004). Photomicrograph in D courtesy of Dr William "Ned" Freidman.*

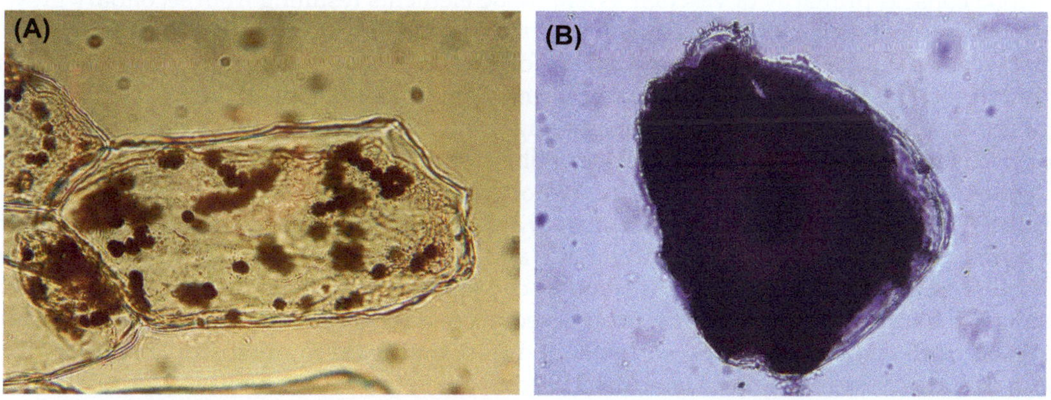

FIGURE 4.17 Starch grains (black spheres) stained with iodide solution in (A) an apple parenchymal cell, 25× and (B) in a potato parenchymal cell. Note that the potato cell contains so much starch that the entire cytoplasm appears black. *Photomicrographs by author.*

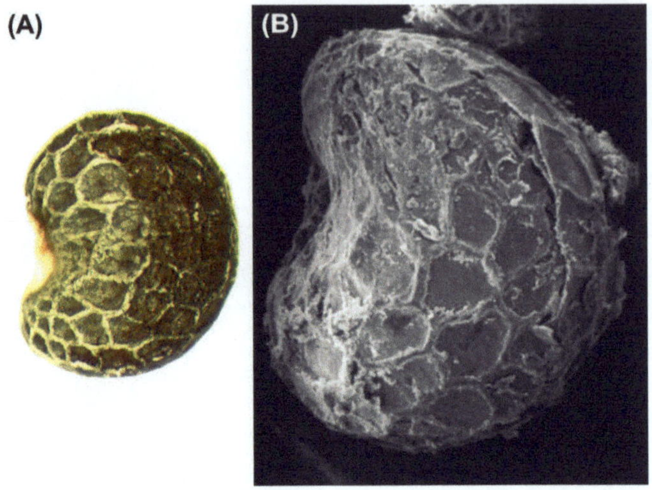

FIGURE 4.18 Poppy seed (see also Figure 3.2). (A) From a fecal sample. (B) Scanning electron micrograph of fresh poppy seed. *Photomicrograph by author; SEM courtesy of Dr. Meredith A. Lane.*

1.3 Fruits and Seeds

Edible fruits and seeds also can be useful forensic tools (Lipscomb and Diggs, 1998). Many fruits have special structures that stick to animals and in turn are transported by them to other locations. These same fruits may adhere to shoes or clothing of suspects and may tie a suspect to a crime scene. Furthermore, humans consume many fruits as well as intentional and inadvertent swallowing of seeds as part of their regular diets, and their identification in GI contents and feces (Figure 4.18) may be of special forensic interest.

1.4 Wood

Wood consists primarily of vascular xylem tissue elements resulting from secondary growth of stems. In older stems, the phloem is found toward the outside and near the surface of the stem, whereas the xylem is located toward the center. The xylem is separated from the phloem by a thin active layer of cells called the **cambium**. Throughout the life of the tree, the cambium produces a small amount of new phloem but huge amounts of xylem cells so that the bulk of the stem is xylem (tracheids and/or vessels) and fibers or what we simply call wood. As the stem of a young tree becomes larger and clearly woody in nature, it is called a trunk. New wood is produced annually. Cells with larger diameters are produced in the spring of the year or at the beginning of the wet season in tropical climates and smaller diameter cells are produced toward the fall or at the beginning of the transition to dry seasons in the tropics. This growth pattern results in the formation of growth rings observed in cross sections of a tree trunk (Figure 4.19).

Wood from conifers (so-called softwoods) differs markedly in microscopic sections from those of eudicot flowering trees (so-called hardwoods) because conifers lack vessels that are prominent in the eudicot stems. It is the pattern formed by the xylem cells that gives different woods their characteristic grain patterns. Since the wood of each tree species can vary somewhat in their grain patterns and, depending on how it is cut, it is often possible to connect broken or sawed pieces of wood or branches together such as in the famous Lindbergh case

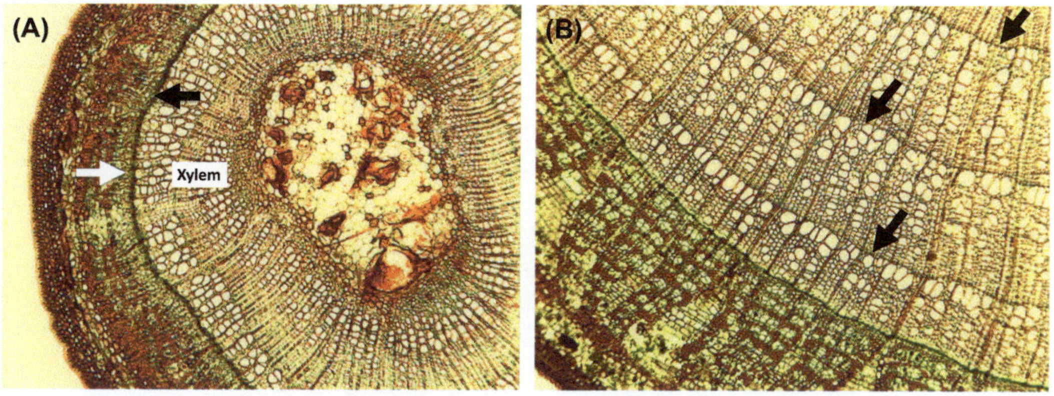

FIGURE 4.19 Cross sections through a woody stem. (A) One-year-old stem. The arrow identifies the location of the cambium. Abundant xylem cells are found toward the center of the stem and a thin layer of phloem cells (P) appears to the outside of the cambium. (B) Three-year-old stem at the same magnification. Note appearance of distinct growth rings (arrows). *Photomicrographs by author.*

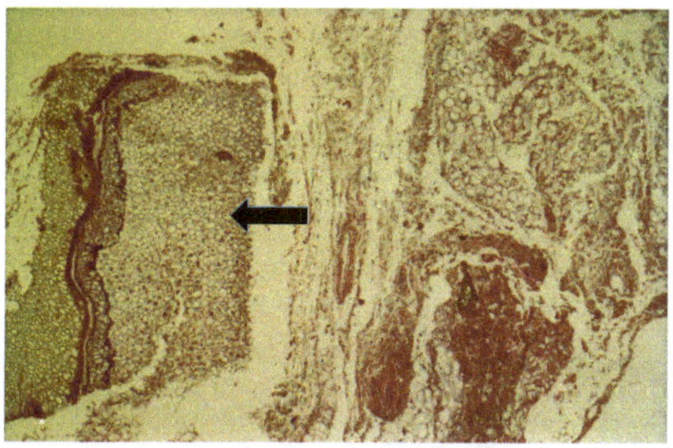

FIGURE 4.20 Wood fragment (arrow) photographed in esophageal tissue from an autopsy of an abused child. *Slide courtesy of Dr. William (Ben) Galloway. Photomicrograph by author.*

(see Chapter 1, p 20) or in the Mirabel case (see Chapter 9, p 124). Toolmarks made on wood can also be matched to a suspect's tool (e.g., see Fisher, 1993).

Identification of wood generally requires a relatively large sample and often can be identified only to genus and not to species (USDA, 2014). For example, you can identify "spruce" but not the kind of spruce or "oak" but not the specific oak. However, knowledge of the locale can narrow the possibilities within a group. There is only one of the red pine group of species native to North America so the field is limited to only one species.

Fragments of wood are sometimes left at crime scenes and sometimes can be matched to their damaged source (see Miller, 1994). Wood may appear in tissues of corpses at autopsy and can be examined microscopically with an aim to identify the source (Figure 4.20).

Because different species produce different sizes and proportions of xylem cell types, we attempted to characterize a number of hardwoods and softwoods by macerating small amounts of wood (see Appendix VI for the procedure) and separating out and measuring the various kinds of xylem cells: ray cells, tracheids, fibers, and vessels. Fibers (sclerenchyma) differ from tracheids in that although both are elongate, narrow cells, the latter exhibit bordered

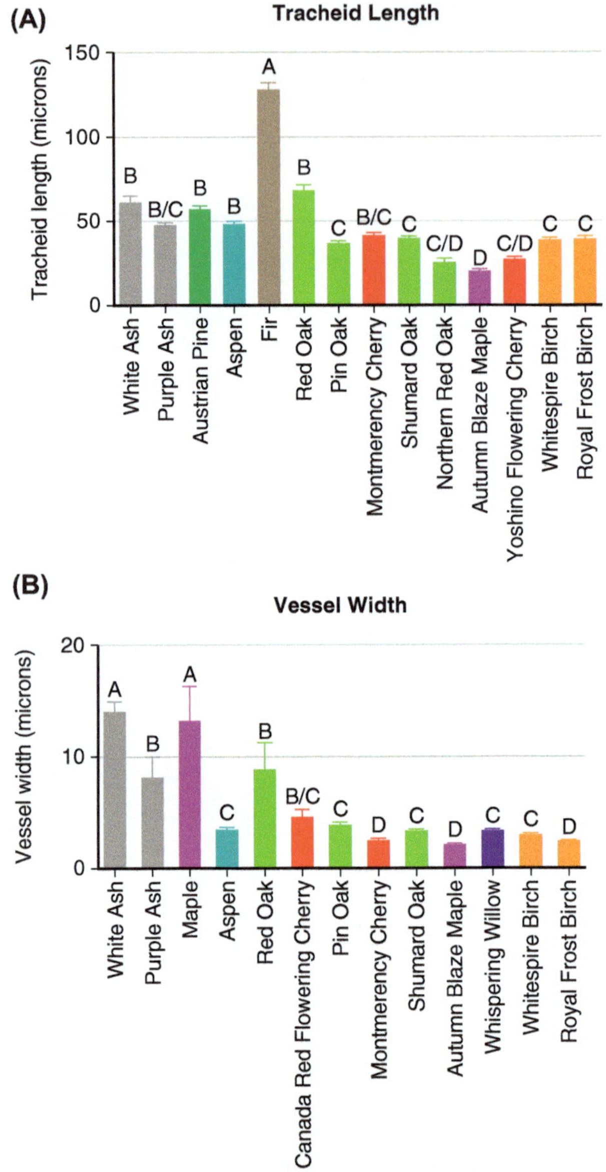

FIGURE 4.21 Measurements of isolated xylem cells following maceration of wood samples. (A) Tracheid length in a conifer and 13 species of angiosperm trees. (B) Vessel widths in 13 species of angiosperm trees. Those with different letters are statistically different ($p < 0.05$). *Norris, D.O., Friedman, W., Bock, Knaub, C. and Kuenning, R, unpublished data.*

pits near each end. Vessels are much greater in diameter than fibers or tracheids and are absent in the softwoods (Figure 4.9). Ray cells are extremely short and rectangular in shape when viewed microscopically as compared to the other xylem cells. Examination of five hardwoods and five softwoods revealed differences in certain measurements among some of the species that might be useful for identification or comparison with an unknown sample (Figure 4.21). More studies are needed in this area.

2. THE HUMAN DIGESTIVE SYSTEM

Information concerning the identification of plant fragments rescued from the **gastrointestinal (GI) tract** can be employed to aid the determination of time of death as well as to connect a suspect to a crime scene. Recognition of the cellular composition and arrangement of those cells in various food species allows for this identification. Since the digestive system typically receives little attention in modern physiology classes, we are providing a brief overview of the human digestive tract and some information on processing of food as a background for the forensic interpretation of plant fragments found in GI contents.

2.1 Overview of Human Digestion and the Digestive Tract

The human digestive system (Figure 4.22) or GI tract consists of the mouth, esophagus, stomach, small intestine, large intestine (colon), and rectum as well as associated glands. Digestion is aided by the secretion of a variety of separate glands (salivary glands; liver and its storage site for bile, the gallbladder; pancreas) as well as glands in the lining of the stomach and small intestine of the digestive tract (Magee and Dalley, 1986).

The process of digestion begins in the mouth where food is chewed and mixed with saliva to form a **bolus**. The enzyme **salivary amylase** (= ptyalin) secreted by the salivary glands begins the process of converting starches to simple sugars. The bolus is swallowed once it achieves proper consistency and travels down the esophagus to the **stomach**. Once in the stomach, the bolus is mixed through the contractions of layers of smooth muscle in

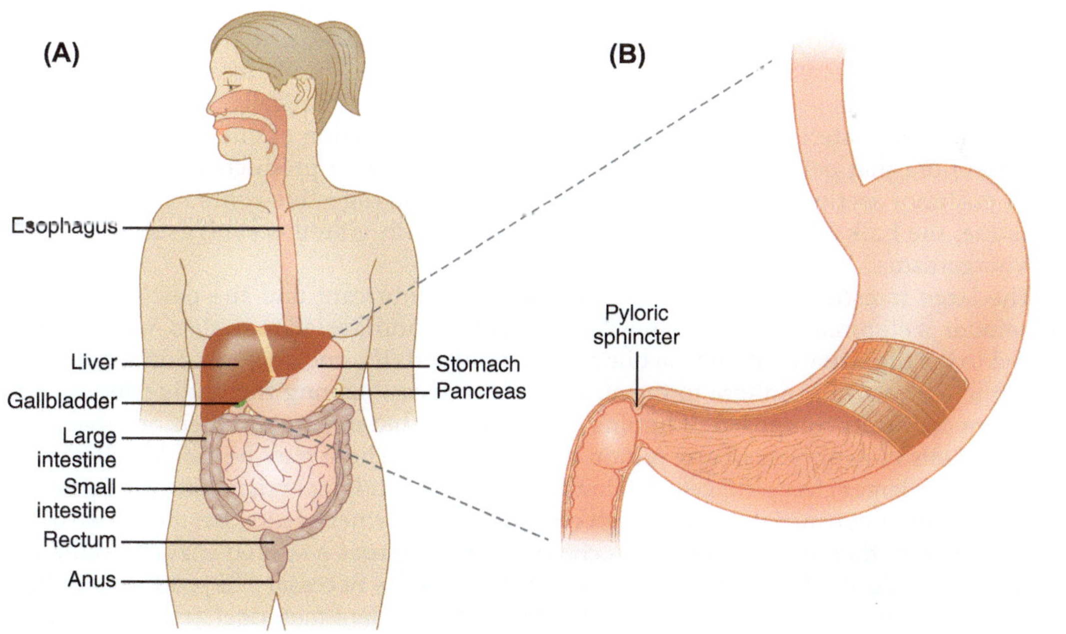

FIGURE 4.22 (A) Human digestive tract. (B) Closing of the pyloric sphincter muscle at death ensures retention of stomach contents as long as the stomach remains intact.

the stomach walls with secretions from the stomach lining. These secretions include **hydrochloric acid** (HCl), a protein-digesting protease known as **pepsin**, and additional water to form **chyme**. The increased acidity inactivates the salivary amylase and stops starch digestion while the food is in the stomach. Pepsin, activated by the HCl, begins to degrade protein into polypeptides. The HCl softens and may even dissolve some materials such as bone particles or other hard items and also kills many bacteria ingested with the food. The stomach lining also secretes mucus to protect the stomach lining from the acid and pepsin mixture.

Once the chyme reaches the proper consistency and acidity, the general assumption is that the **pyloric sphincter** (Figure 4.22), a muscle at the posterior or pyloric end of the stomach, relaxes and the acidic chyme is squirted into the **duodenum** of the **small intestine**. However, removal of the diseased pyloric region apparently has no effect on gastric emptying, and it is thought that it may only regulate passage of solid items into the duodenum (Magee and Dalley, 1986). Typically, solid materials that are passed on to the duodenum are limited to approximately 2 mm in diameter, although passage of larger nonfood items has been observed. Activities of the stomach can be slowed by hormones secreted from the duodenum if the fat content or acidity of the stomach is too high (as detected in the chyme entering the duodenum).

Additional enzymes are added to the contents of the duodenum by the pancreas and glands in the lining of the duodenum. Basic fluids are also secreted from the pancreas to neutralize the acid in the chyme. If the chyme contains significant amounts of fat and/or oils, a hormone released by the duodenum stimulates the gallbladder to contract and release bile that enters the duodenum via the bile duct. Bile separates the fat into tiny droplets that can be more readily accessed by fat-digesting enzymes (**lipase**). The mixture of **pancreatic amylase**, additional **proteases, peptidases** (that separate polypeptides into amino acids), and lipase completes the process of digesting food items to simple sugars (e.g., glucose and other simple sugars from starches and glycogen), amino acids (from proteins), fatty acids, and glycerol (from fats and oils) most of which are then absorbed into our bodies via the duodenum. Nucleic acids such as DNA and ribonucleic acids (RNAs) also are degraded enzymatically, and the products are absorbed. Food material continues to be processed by the action of our enzymes and by the bacterial flora that permanently resides there. Additional end products are absorbed in the more posterior sections of the small intestine. When the digestive contents reach the **large intestine**, the bulk of the water is reabsorbed along with vitamins synthesized by resident microorganisms.

The large intestine secretes mucus that acts as a lubricant and the partially solidified remainder is eliminated as **feces**, consisting largely of undigested materials, waste products added to the digestive contents via bile or intestinal activity, and bacteria. Some human cells from the GI tract may adhere to the surface of the feces. Among the undigested materials are plant cell walls composed of lignins and cellulose that may retain their original shapes.

Plant cells may appear in any region of the GI tract. There may be single cells, isolated trichomes, or larger clumps of cells retaining some of their original orientations as found in the plant prior to its consumption. Although cellulose is a polymer of glucose, it is constructed in a different way than is glycogen or starch and hence requires a special enzyme, cellulase, to break it apart. Animals (including humans) lack the genes necessary to produce a cellulose-digesting enzyme, and those animals that exclusively eat plant material (termites, cattle, deer, etc.) rely on microorganisms (e.g., bacteria or other protists) living commensally in their guts

TABLE 4.3 Estimated Times of Food Processing for a Full Meal in Various Compartments of the Human Digestive Tract[a]

Region	Time (h)
Mouth	<0.03
Stomach	2–6
Small intestine	2–8
Large intestine	6–9
Total time for a single meal[b]	10–23

[a] *These data were generated from an analysis of the published literature and were originally presented by Bock and Norris at the annual meeting of The American Academy of Forensic Sciences in 2001.*
[b] *Unusual rates from mouth to anus of 31–118 h (almost 5 days) have been reported.*

to digest the cellulose. Thus, plant material isolated from any part of the human digestive tract or from feces can provide evidence of what has been eaten recently.

As indicated above, the digestive process, and hence the time required for food processing, differs in the various regions of the digestive tract (Table 4.3). Generally, food will be swallowed after only a brief presence (typically, less than 1 min) in the mouth with a minimal exposure to salivary amylase. After a full meal, food will remain in the stomach of a healthy adult 2–6 h, depending on a variety of variables (see ahead). After death, the pyloric valve closes and what is in the stomach at death stays there as long as the stomach remains intact. Digestion products are not absorbed while in the stomach although most drugs are absorbed here and can influence digestive processes (see ahead).

Once chyme enters the small intestine, it will migrate back and forth within the small intestine for 2–8 h before passing on to the large intestine where it will remain for an additional 6–9 h. Most of the digestive products obtained from the meal are absorbed by the small intestine into the body. The large intestine mainly reabsorbs water and vitamins produced by the intestinal flora. Thus, the total digestive time to covert a single meal to feces in a normal person can vary from 10 to 23 h. Since most healthy humans defecate once or twice per day, a fecal sample routinely may contain undigested plant material from one to three or possibly four meals depending on the frequency of consumption and defecation.

2.2 Experiments on Gastric Evacuation

The earliest studies on gastric emptying (evacuation) were conducted in the early 1800s by **Dr William Beaumont**, an army surgeon at Ft Mackinac, MI (Edwards, 2010). A young French trapper named **Alexis St. Martin** was accidently wounded with a shotgun blast at close range that left a gaping wound in his side and a hole in the wall of his stomach (Figure 4.23). Under Beaumont's care, the trapper miraculously survived. In fact, St. Martin made a complete recovery under Beaumont's continuing care save for one anomaly: during the healing process, the wall of his stomach and the body wall fused together leaving a small channel that connected the inside of his stomach to the outside world. Over a period of years, Beaumont was able to observe the process of gastric digestion in

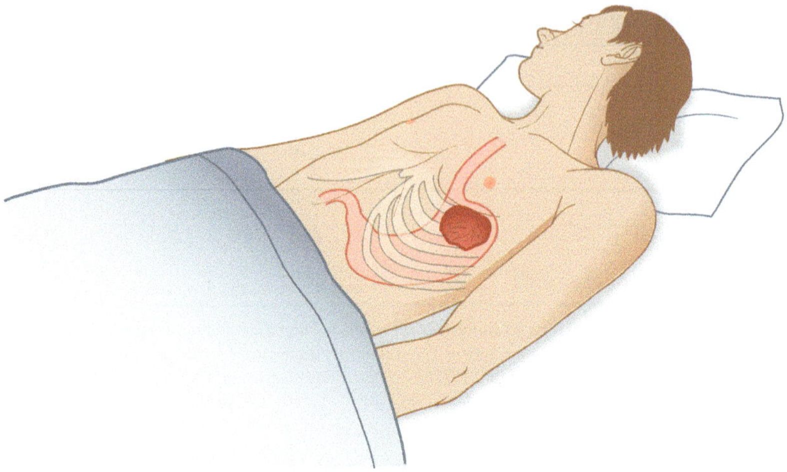

FIGURE 4.23 Alexis St. Martin. View of his wound. He survived due to Beaumont's care, but the wall of the stomach fused with the body wall leaving an opening directly into the stomach.

St. Martin's stomach as well as discover that the acid produced by the stomach was HCl. These early studies led to the later development of experimental "windows" (in domestic animals such as sheep and cattle) called gastric fistulas that allowed for more sophisticated studies of gastric digestion. Eventually, Beaumont published his observations in a book (Beaumont, 1833). Some of Beaumont's original observations on digestion within St. Martin's stomach are reproduced in Box 4.1.

Modern studies of human gastric evacuation use a variety of more sophisticated, noninvasive approaches to calculate the time it takes for the stomach to evacuate one-half of a test meal (T=½). One method involves **scintigraphy** that measures the passage of food labeled with radioactive technetium (^{99}Tc). Another measures the ratio of exhaled $^{13}CO_2$ to $^{12}CO_2$ following the ingestion of ^{13}C-octanoic acid. ^{13}C is a nonradioactive (i.e., stable) isotope of carbon (^{12}C). Such studies indicate a T½ in healthy volunteers from about 60 to 100 min with only about 2% of a meal remaining 4 h after a moderate meal (see Table 4.5). Interestingly, these modern studies demonstrate rates of gastric evacuation similar to those described by Beaumont for St. Martin nearly two centuries earlier.

If the composition of a person's last meal is known and it matches the contents found in his/her stomach after death, one can make a reasonably close estimate from the volume and condition of the stomach contents as to when the person died. However, many variables can affect the rate of gastric evacuation and hence the estimate of the **postmortem interval** (PMI). The rate of gastric evacuation can be affected by the composition of the meal, the temperature of the meal, and the size of the meal as well as by consumption of alcohol, drugs, or water prior to or during the meal. (For example, consumption of ice-cold or alcoholic beverages typically slows down the process (Table 4.4 and 4.5).) Activity after consuming a meal actually accelerates gastric evacuation, allowing less

BOX 4.1 BEAUMONT'S OBSERVATIONS ON ST. MARTIN

Excerpts from Beaumont's book "Experiments and Observations on the Gastric Juice and the Physiology of Digestion" published in 1833:

William Beaumont maintained a connection with his former patient St. Martin and continued his experiments over a period of years. He would provide St. Martin a meal and then collect samples through the opening in his stomach (Figure 4.26). He took detailed notes on his observations, some of which are reproduced here.

April 7

At 8 a.m.—(St. Martin) breakfasted on *three hard boiled eggs*, *pancakes*, and *coffee*.

At 8:30 a.m.—examined stomach—found a heterogenous mixture of several articles eaten, slightly digested.

At 8:45 a.m.—examined again—found contents reduced in quantity, and changed in quality—about half digested.

At 10:15 a.m.—no part of the breakfast remained in the stomach.

At 11:15 a.m.—he ate *two roasted eggs* and *three ripe apples*. In 30 min, examined stomach—found a heterogenous mixture, in an incipient stage of digestion.

At 12:15 p.m.—examined again—found the stomach clear; no vestige of apples or eggs.

At 2 p.m.—dined on *roasted pig* and *vegetables*.

At 3 p.m.—examined, and found it about half chymified.

At 4 p.m.—very little remained in the stomach.

At 4:30 p.m.—nothing remained but a very little gastric juice.

April 9

At 3 p.m.—he dined on *boiled dried codfish*, *potatoes*, *parsnips*, *bread*, and *drawn butter*.

At 3:30 p.m.—examined, and took out a portion, about half digested; the potatoes the least so of any part of the dinner. The fish was broken down into small filaments; the bread and parsnips were not to be distinguished.

At 4 p.m.—examined another portion. Digestion had regularly advanced. Very few particles of fish remained entire. Some of the potatoes were distinctly to be seen.

At 4:30 p.m.—took out, and examined another portion—all completely chymified.

At 5 p.m.—stomach empty.

FIGURE 4.26 Painting depicts Dr William Beaumont collecting a gastric sample from Alexis St. Martin. *Source: Mackinac State Historic Parks Collection.*

TABLE 4.4 Some Factors Affecting Gastric Evacuation Time

Factor	Effects	Source
Volume of meal	Larger meals took longer	Noakes et al. (1991) , Horowitz et al. (1994)
Composition of Meal		
Solid/liquid ratio	Higher solid portion slowed	Holt et al. (1986) , Horowitz et al. (1994)
High caloric content	Slowed	Hunt and Knox (1968), Horowitz et al. (1994)
High fat content	Slowed	Stacher et al. (1991)
High carbohydrate	Slowed	Rehrer et al. (1989)
Ethyl alcohol	Slowed	Barboriak and Meade (1970)
Activity	Accelerated	Knight et al. (1997)
Sex	Males emptied meal faster than females	Hermanson and Silversston (1996)
Age	Slowed	O'Donovan et al. (2005)
Race	Faster in nonsmoking Hispanics vs nonsmoking non-Hispanic whites	Schwartz et al. (1995)
Fear	Slowed and may even be stopped	
Obesity	Slowed	Horner et al. (2014) , Omür et al. (2014)
Pathologies		
Diabetes mellitus	Slowed	Horowitz et al. (1996)
Dyspepsia	Slowed	Bromer et al. (2002)
Duodenal ulcer	No effect	Holt et al. (1986)
Cirrhosis of liver	Slowed	Schoonjans et al. (2002)
Gastroparesis	Slowed	Hasler et al. (2008)
Surgery		
Gastroenterostomy	Accelerated	Magee and Dalley (1986)
Radical antral resection	Accelerated	Michalsky et al. (2013)

time for proper processing of the food (Figure 4.24). Sex also influences the rate of gastric evacuation with adult males exhibiting faster evacuation rates ($T½ = 71$ min) than females ($T½ = 102$ min) of comparable size (Figure 4.25). Numerous pathological conditions can alter the time necessary to process a meal so it is imperative to have some knowledge of the victim's health history. Although duodenal ulcers do not affect gastric emptying, obesity slows the process significantly. Psychological factors such as fear also can alter the digestive process. Nevertheless, a reasonable estimate of the PMI can be made from a match of food items with the last known meal (see Chapter 5 for some specific examples).

TABLE 4.5 Some Gastric Half-Emptying Times (T½) for Adult Humans

References	T½ (min)	Meal composition	Sex	Age (year)	Body mass index (BMI)
O'Donovan et al. (2005)					
Younger group	~120	75 g glucose, 600 ml H_2O	7M:3F	24.5 ± 2.2	21.8 ± 0.6
Older group	~150		4M:4F	73.5 ± 1.7	24.1 ± 0.8
Bromer et al. (2002)		One muffin			
	64 ± 17	Solids	6F:4M	34	
	55 ± 27	Liquid	6F:4M	34	
Barboriak and Meade (1970)					
No whiskey	105 ± 19	Eggs, toast	M (8)	25–75	
After 4 oz whiskey	204[a]	Butter, coffee, "Half & Half"			
Schwartz et al. (1995)					
Mexican American	49.9 ± 4.7	450 ml H_2O, 50 g glucose	M (18)	31.1 ± 1.5	26.6 ± 0.7
	63.1 ± 5.3		F (14)	30.1 ± 1.7	22.9 ± 0.8
Non-Hispanic white	59.6 ± 4.6		M (18)	33.1 ± 1.5	25.3 ± 0.7
	73.2 ± 5.4		F (14)	30.0 ± 1.8	23.7 ± 0.7
Hermanson and Silvertsson (1996)					
	111.2 ± 34.5	Swedish pancakes, 20 g jam, no liquid	M	20–29	21.7 ± 3.3
	158.2 ± 24		F	20–29	21.1 ± 2.0
Knight et al. (1997)					
	71	Two egg sandwich on white bread, 300 ml H_2O	M (13)	27.5 ± 1.7	
	102		F (9)	27.9 ± 2.2	
Moore et al. (1990)					
Stand at rest	72.6 ± 7.6	150 g beef stew, 150 g orange juice	M(10)	22–44	
Walk at 3.2 km/h	44.5 ± 3.9				
Walk at 6.4 km/h	32.9 ± 1.9				
Horner et al. (2014)					
Obese subjects	179 ± 15	Standard pancakes (400 kcal), 250 ml H_2O	M (15)		30.3 ± 4.9

[a] *Based on increased time of 99 ± 32 min longer after ingestion of alcohol 15 min prior to the meal since there was considerable variation among the subjects for the T½ of the control meal.*

2.3 Common Food Plants

Simpson and Ogorzaly wrote (1995, p. 1), "It has been estimated that about 3,000 species of plants have been used as foods by human beings and that about 200 have been domesticated as food crops. Tragically, the number of species used has been decreased rather than increased within historic times."

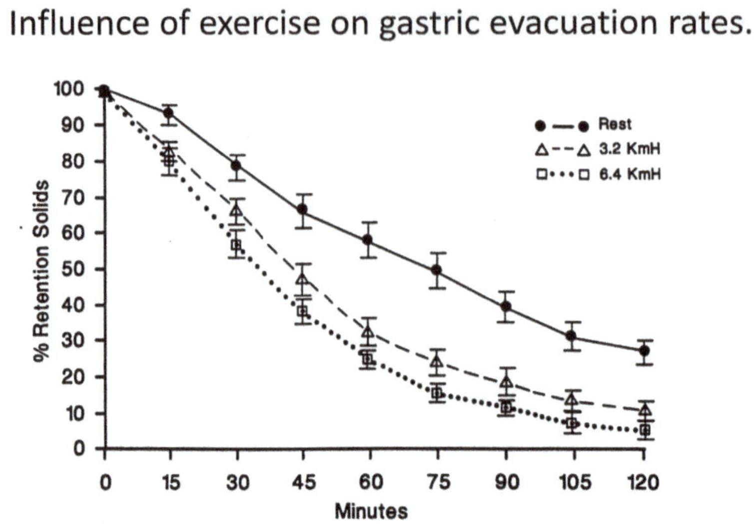

FIGURE 4.24 Effects of exercise on gastric evacuation rates in men. Walking at a rate of 3.2 or 6.4 km/h increased the T½ (44.5 ± 3.9 and 32.9 ± 1.9 min, respectively) as compared to resting T½ (72.6 ± 7.6 min) following consumption of a meal consisting of beef stew. *Adapted with permission from Moore et al. (1990).*

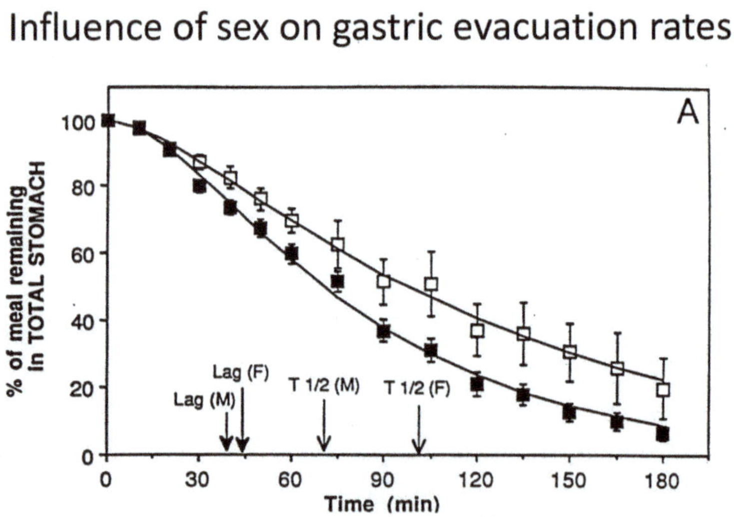

FIGURE 4.25 Gastric evacuation rates in men compared to women of comparable size and conditioning. T½ occurred significantly sooner in men than in women (71 and 102 min respectively; $p = 0.021$). *Adapted with permission from Knight et al. (1997).*

In the contemporary diet of modern North American humans, only about 70 plant food species are used. This low number includes regional diets and special foods. Our estimate is based on informal surveys of grocery stores, farmers' markets, and cookbooks. Local grocery stores often feature several versions of a species. For example, there may be seven or eight varieties of lettuce, but from the cell composition perspective, lettuce is lettuce, onions are onions, apples are apples, potatoes are potatoes. All the citrus fruits (e.g., oranges, lemons, grapefruit, and limes) may be lumped together based on the structure of their edible portions. Similarly, Navy beans, Great Northern beans, red kidney beans, pinto beans, Anasazi beans, and black (turtle) beans that are all varieties of the same species, *Phaseolus vulgaris* appear similar under the microscope. As mentioned earlier, kale, cabbage, cauliflower, broccoli, kohlrabi, and Brussels sprouts are the same species. Several types of squash including zucchini and pumpkin are also varieties of a single species (Table 4.2).

3. PLANT CELLS AND TIME OF DEATH

In spite of the tremendous improvements in forensic techniques in recent years, a perfect technique for determination of time of death or PMI has remained illusive (see Sachs, 2001). Yet, determining when a victim died is of paramount importance in a homicide case as it indicates which suspects might have had the opportunity to commit the crime. If a body is recovered before decomposition is too advanced, commonly employed methods, such as cooling of the body, appearance and loss of rigor mortis, and timing of invasive insect development are all affected by a number of variables resulting in confidence limits of hours or days. Consequently, it is often necessary to combine a variety of methods to estimate a window as small as possible for when the victim might have died.

Because the pyloric sphincter muscle closes down upon death, any food remaining in the stomach will provide clues to the last meal. When the body of a woman missing since November was discovered the following spring as the snow melted on a mountain pass in Colorado, her stomach was still intact and it was possible for us to estimate the contents of her last meal based on the plant material present.

Examination of stomach contents can be a useful method for determining time of death. As indicated above (see Table 4.3), a full meal resides in the stomach for 2–6 h depending on a number of variables (Table 4.5). If investigators have information concerning the contents of the last known meal the victim consumed that can be verified by witnesses, examination of the stomach or intestinal contents may be useful for estimating time of death. For example, if the victim's last known meal contained corn, onions, lettuce, and tomato but the stomach contained green beans, potatoes, cabbage, and spinach, the victim must have consumed at least one additional meal before death. On the other hand, if the contents matched the last known meal, the window for time of death could be limited to a few hours.

4. COLLECTION AND SAMPLING METHODS FOR DIGESTIVE TRACT MATERIALS

4.1 Collection of Samples

4.1.1 Gastric and Intestinal Material

Stomach or intestinal contents can be collected at autopsy. If it is not feasible to retain the entire sample, it can be thoroughly mixed and a portion saved for later analysis or several subportions can be selected to represent the entire contents. This is especially important if the meal is in the early stages of digestion and there are many clumps of solid or semisolid material present. For intestinal contents, several samples should be preserved that reflect the various regions. A preservative should be added or the sample should be frozen to prevent further changes. Although the structure of plant cell walls (cellulose or silica) are not affected by drying, freezing, cooking, or gastric enzymes, internal cellular components such as crystals or organelles (e.g., plastids or vacuoles) might be affected, and hence it is important to prevent any further changes.

4.1.2 Collection of Vomitus or Fecal Matter

Fresh fecal matter or vomitus collected at a crime scene or at autopsy may be frozen or stored in a preservative for analysis of plant fragments. The Technical Working Group on Biological Preservation (2013) suggests freezing of feces is the best general method for short- or long-term preservation. The Working Group did not address vomitus as a biological sample. Dry feces or vomit can be stored in labeled paper, glass, or plastic containers.

Special care should be taken for fecal material or vomitus that may be adhered to clothing because it may slough off during handling. Swatches of clothing exhibiting the biological material should be carefully cut from the clothing and placed in sealed containers (paper, glass, or plastic containers are suitable) to ensure the plant material will not be lost. When the stains are only 1–3 mm across, it is especially important to provide as many samples of stained clothing as possible since by chance small stains may not be representative of the original mass (see Norris and Bock, 2000). Since material adhering to clothing is usually dried, no preservatives are necessary. If the material is damp, it should be dried before sealing it in the storage container.

5. PROCESSING OF FORENSIC SAMPLES USING PLANT ANATOMY

Procedures for processing samples taken directly from the GI tract at autopsy or at a crime scene are separated here from those used for examining vomitus or feces from clothing of a suspect.

5.1 Processing of Stomach or Intestinal Samples

Stomach or intestinal samples should be disinfected prior to examination. If not already treated with formalin, sufficient formalin should be added to equal 5–10%. This will denature all potentially infective agents without harming the plant material. Frozen samples can be defrosted by immersion in a 5–10% formalin solution. Similarly, dry samples can be rehydrated in dilute formalin. 100% formalin is a saturated solution of water achieving a final concentration of 37% formaldehyde.[4]

5.2 Preparation of Known Plants for Identification

There are numerous guides available for the identification of flowers, fungi, ferns, and trees and other plants, yet there are no good guides for identifying the plants we eat, especially from an anatomical point of view. Andrew and Kate Winton published a monumental four-volume work at the beginning of the twentieth century on *The Structure and Composition of Foods*. Volume 2 of that work focused on "vegetables, legumes, fruits" (Winton and Winton, 1935). Of the 243 food plants they discuss in this volume, it is unlikely any of you has eaten 104 of them (e.g., New Zealand spinach). You might find it hard to conceive of any human eating some of them (e.g., great burdock roots and kudzu). Although detailed anatomical drawings do not accompany the discussions of many of these plants, some of the drawings are very detailed and useful (e.g., blueberry; see Figure 4.27). We published a black and white photographic atlas of some of the anatomical features of 42 common edible food plants, but it is out of print (Bock et al., 1988). Additionally, we are providing a more extensive online atlas of food plants that have useful anatomical features for identification (http://booksite.elsevier.com/9780128014752). The best reference collection, of course, is the one you prepare yourself from raw and/or cooked plants relevant to each case.

5.2.1 *Developing a Reference Collection*

The local grocery store is a great source for common food plants. Since one needs very little material, we have found produce managers in grocery stores willing to provide scraps of fresh plant foods once we explain why we need them. Canned products often can substitute for fresh since processing does not affect the plant cell anatomy (although it may affect some inclusions). A permanent collection of commonly consumed plants can be a useful reference tool. These can be immersed in 70% ethanol and stored in tightly sealed containers. A set of permanent microscope slides also may be prepared for reference (see Appendix I). A photographic atlas of cell types from each food plant also may be useful.

A listing of materials needed for preparing samples for microscopic examination is provided in Appendix I. A known sample of each plant can be prepared separately by mincing the plant with a clean, single-edged razor blade or *gently* grinding it with a mortar and pestle. Heating or cooking the plant may soften it and make maceration easier (for example, carrots). Caution: vigorous grinding usually destroys features necessary for identification. It is very important to always use clean utensils when preparing each plant type so as not to contaminate one plant with cells or fragments from a different species during preparation.

Generally, one is not identifying single cells (no one chews their food that well) but rather fragments of plant tissue with cells exhibiting particular arrangements with each other characteristic of that plant. A small subsample of the thinly minced or lightly ground plant can be placed on a clean glass microscope slide, a few drops of water placed over the sample, and a glass coverslip attached. This is known as a wet mount. To affix the coverslip to the

[4]Formaldehyde is a highly toxic chemical to all animals and should be used with caution. It is also known to be carcinogenic.

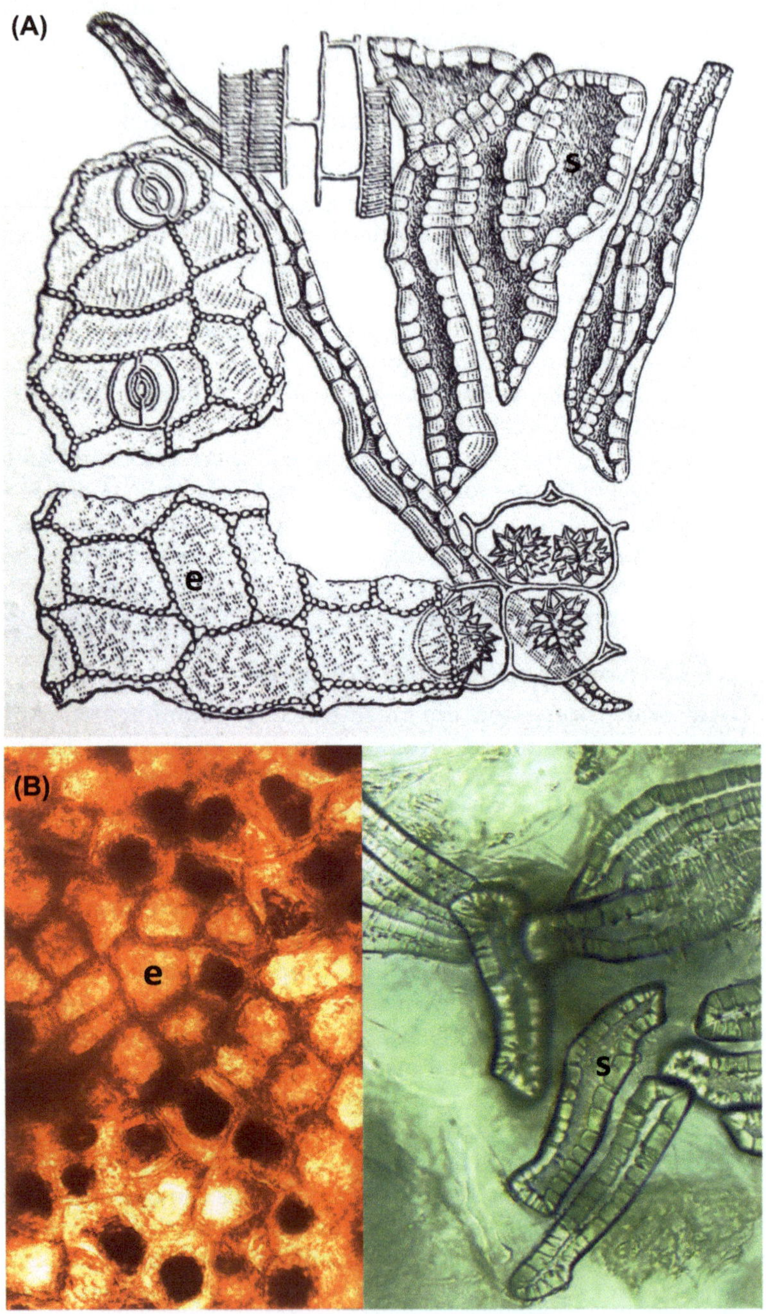

FIGURE 4.27 (A) Seed coat epidermis (e) and stone cells (s) or sclereids (collenchyma) of blueberry, *Vaccinium* spp. *(Reprinted with permission from Winton, A.L., Winton, K.B., 1935. The Structure and Composition of Foods. Vol. 2. Vegetables, Legumes, Fruits. John Wiley & Sons, New York.)* (B) Photomicrograph by author from a smashed fresh blueberry showing seed coat epidermis (e) and sclereids (s), 25×.

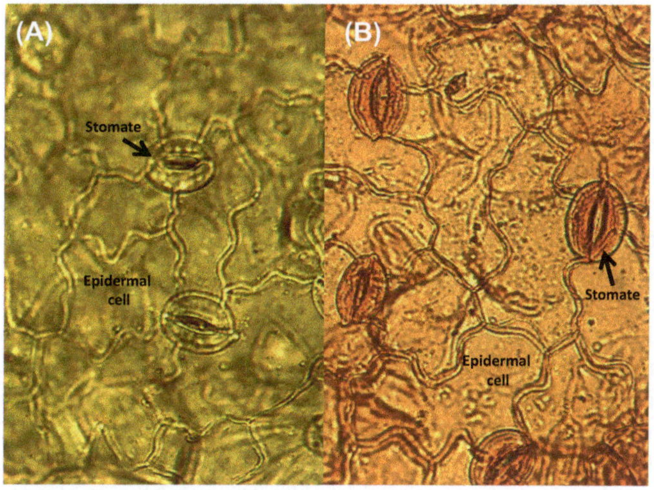

FIGURE 4.28) Comparison of epidermis of fresh spinach leaf (A) unstained and (B) stained with safranin. *Photomicrographs by author.*

slide, touch the edge of the water droplet with the coverslip and lower it slowly to cover the sample by using a dissecting needle. This will help prevent formation of air bubbles that can make viewing the material difficult. Sometimes it is best to further macerate the tissue on the slide with a clean razor blade or with dissecting needles prior to adding the coverslip as it is important to get the material thin enough that light will readily pass through it. A dye such as safranin O or toluidine blue (see Appendix IV) can be added to increase the contrast and provide better detail (Figure 4.28). Sometimes it is useful to examine and document the fresh unstained material, then carefully remove the coverslip, making sure that all of the material remains on the slide, then add dye to the same slide and reaffix the coverslip. Addition of more water may be necessary along with the dye to obtain a bubble-free preparation. One should practice doing this with noncritical samples to develop some skill at this process before examining samples from cases. If the plant fragments in the resulting preparation are too thick so that cellular details are obscured, carefully remove the coverslip and again mince the fragments as above. Add some more water if necessary and reapply the coverslip. Repeat if necessary. Alternatively, start with fresh tissue and prepare it more thoroughly.

5.3 Identification and Documentation of Known Plants and GI Tract Contents

Identification of the plant can be made with the aid of a compound microscope using 10× or 25× objectives and comparing the unknown to a collection of known samples prepared in a similar manner. A 4× objective may be useful to get an overview of what is on the slide. Sometimes 40× is useful especially if dealing with small cells such as diatoms, but it is usually unnecessary due to the relatively large size of seed plant cells as compared to those of animals. A compound microscope with a phototube adapter equipped with a digital camera paired to appropriate computer software should be used to document what is seen on the wet mount slide since this type of slide is not permanent. Digital photographs of the measurement scale on a stage micrometer (see Appendix III) should be made using each microscope objective for later measurement of the plant material photographed. All plant fragments seen should be documented with photographs, noting the magnification of the microscope objective used. These can be saved in jpg format on the computer for future reference. Backups should also be saved. It is a good practice to print copies of all in jpg format and save them in a paper file by case number.

We have found that this method of preparing wet mounts aided by visual digital recording is the most efficient and rapid procedure for examining GI contents. Alternatively, permanent slides can be made initially and examined and documented later (see Appendix I). However, making permanent slides requires considerably more time. It is also possible to make a permanent slide of the material on the wet mount slide after examination and documentation (see also Appendix I).

Once familiar with possible known plants, several random samples of a thoroughly mixed GI tract sample can be examined as wet mounts in a similar manner. We have found it necessary to dilute the forensic sample (especially true for fecal material) so that the concentration of background materials is less, thus allowing more light to pass through the preparation and yielding better photographs. If the composition of a last meal is known, this will tell you what plants should be present and what ones should not be present. All plant fragments seen should be documented with photographs, noting the magnification of the microscope objective used. One should continue to examine subsamples of the GI contents until no new items are found (typically fewer than 10 samples if the original sample was well mixed). Before they are macerated for microscopic examination, larger items (e.g., corn kernels, peas, etc.) should be removed prior to mixing and photographed using a close-up camera or a dissecting microscope with a digital camera. Sometimes, other types of cells may be present in a sample (e.g., muscle cells from meat may be present and still recognizable in a stomach sample), as well as seeds or even insect body parts. If the victim was recovered from water, diatoms, or other aquatic plants or animals may be present in the stomach. All of these items should be documented with photographs.

5.4 Processing of a Fecal Sample

Fecal samples must be disinfected prior to analysis. If the sample is frozen, it should be thawed in 10% formalin. If dried, it should be rehydrated in 10% formalin. If the sample is fairly large, subsamples should be prepared for microscopic analysis as described for GI tract samples. Generally, fecal material already is well mixed, but it may contain material from more than one meal and larger pieces (including seeds) may not be uniformly distributed. Hence, it is best to take several samples from different regions to better characterize the entire mass. Solid items such as seeds or larger plant fragments should be separated for examination with a dissecting microscope. (We have provided photographs of commonly swallowed seeds on the Web site.) Dilution of the original samples prior to slide preparation is very important. Slides can be prepared and documented as described for GI tract contents. See also Norris and Bock (2001).

5.5 Processing Dried Fecal or Vomitus Material Adhering to Clothing

Fresh feces or vomitus can be detected by their characteristic odors, but in some instances it may be useful to employ a chemical test to verify that a stain is definitely feces or vomitus. Established techniques are available that rely on the presence of a bile pigment, urobilinogen, in feces and the presence of gastric fluid for vomitus (see Appendix V).

Fecal material or vomitus adhering to clothing can easily be sloughed off or lost during handling or packaging. Minimally, the piece of clothing should be placed in a sealable plastic bag so that if the fecal or vomitus material comes free, it will be retained in the packaging. Alternatively, the sections of clothing with the stains can be carefully cut out and each piece stored separately in a sealed container. The latter approach is preferable.

The dry fecal or vomitus material attached to the clothing should be carefully scraped from the stain with a clean razor blade onto clean paper and then placed into a prelabeled vial. The storage container should be checked for material that may have sloughed off the clothing and should be processed separately. A small amount of 10% formalin can be added to the vial to disinfect and rehydrate the material prior to examination. The aqueous formalin solution will disinfect, rehydrate, and preserve the sample indefinitely. A sample from the mixture in the vial should be placed on a slide, a few drops of water added, and a coverslip applied as described for gastric contents above. Staining with safranin O will help identify the plant fragments as other material present will not take up much if any of the dye.

If the biological stains are very small, it is important to sample several stains to gain a composite picture of everything that might have been present before comparing it to a sample found at the crime scene or to other stained garments. If possible, separate stains should be examined until no new items are found.

It is important to note that many small fragments (e.g., single cells, dissociated trichomes, crystals, phytoliths, etc.) may be present in fecal samples that you may not be able to identify as to their species source. However, everything present should be documented including its relative frequency in each source as unrelated fecal samples may contain different fragments or the same fragments in different frequencies. A case for identity can be made using the inclusion and abundance of unidentified fragments in addition to those that are clearly identifiable as to their plant source (Norris and Bock, 2000).

In order to establish a clear link between two samples of fecal material, it is best to compare the test samples to one or more unrelated fecal samples. It is relatively easy to obtain daily fecal samples from volunteers to establish a reference bank of unrelated samples.

Similarly, vomit contents can be compared to a known or suspected last meal.

6. THE ROLE OF STATISTICS IN EVALUATING PLANT CELLS IN DIGESTIVE CONTENTS

When people think of scientific data, they automatically associate data with statistical analyses that provide estimates of probability, p. A p-value of less than 0.05 has been adopted as a standard by the scientific community. Many assume this means that if you were to repeat the same study, the probability is that the results will be the same more than 95% of the time. As discussed previously, this is a common misunderstanding and actually is only a measure of acceptance or rejection of the null hypothesis (see Chapter 2, pp 28-29).

Using plant anatomy (cellular structure and conformation) to identify a particular plant generally does not lend itself to statistical comparisons because the identification process simply indicates "yes, it is A" or "no, it is not A." However, in some instances it is not possible to identify a distinct plant, hence "it could be A but it is definitely not B or C," for example. Even so, the presence of a particular plant structure or cell might be useful when comparing two sources of material to see if they are alike in composition (e.g., two fecal samples) even though the species from which it came is not known for certain.

However, when making size comparisons of cells to distinguish two plant species from one another, a sufficient number of cells can be measured by determining widths, lengths, and/or areas. These parameters can be compared statistically using simple t-tests (comparing only two species) or analyses of variance (three or more species; see Figure 4.20).

Each species of plant or plant tissue we consume as food is composed of similar cell types arranged in distinctive patterns. Because of the tremendous agricultural modifications of food plants through selective breeding, food plant cells recovered from digestive contents are often not indicative of a particular variety of the food plant. As indicated previously, several varieties of onion, apple, lettuce, or beans will appear to be the same microscopically even though when prior to consumption they were dissimilar to the naked eye (e.g., Navy vs pinto beans; Macintosh vs a red delicious apple; cauliflower vs broccoli; acorn squash vs pumpkin).

7. SUMMARY

In summary, the microscopic or gross identification of plant food species relies on established principles of observational science rather than on statistical analyses. Frequently, unknown plants in forensic samples are identified by a simple microscopic comparison to known plants. It is important to stress the use of repeated subsamples as mentioned previously to assure a thorough analysis was done. Reliance on a single slide preparation should be discouraged.

Statistics can be useful in laboratory studies for making finer distinctions not possible with simple visual inspection. For example, in order to determine whether epidermal cells of a similar conformation were from garlic or onion, we measured the lengths and widths of many cells and compared their average size statistically. The garlic epidermal cells were significantly smaller than the onion epidermal cells ($p < 0.01$).

References

Barboriak, J.J., Meade, R.C., 1970. Effect of alcohol on gastric emptying in man. The American Journal of Clinical Nutrition 23, 1151–1153.

Beaumont, W., 1833. Experiments and Observations on the Gastric Juice and the Physiology of Digestion. Plattsburgh.

Bock, J.B., Lane, M.A., Norris, D.O., 1988. Identifying Plant Food Cells in Gastric Contents for Use in Forensic Investigations: A Laboratory Manual. U.S. Department of Justice, National Institute of Justice.

Bromer, M.Q., Kantor, S.B., Wagner, D.A., Knight, L.C., Maurer, A.H., Parkman, H.P., 2002. Simultaneous measurement of gastric emptying with a simple muffin meal using [^{13}C]octanate breath test and scintigraphy in normal subjects and patients with dyspeptic symptoms. Digestive Disease and Sciences 47, 1657–1663.

Edwards, L., 2010. The gruesome medical breakthrough of Dr. William Beaumont on Mackinac Island. http://mynorth.com/2010/05/the-gruesome-medical-breakthrough-of-dr-william-beaumont-on-mackinac-island/.

Fisher, B.A.J., 1993. Techniques of Crime Scene Investigation, fifth ed. CRC Press, pp. 191–194.

Hart, T., 2015. Phytoliths: the storytelling stones inside plants. American Scientist 103, 136–143.

Hasler, W.L., Coleski, R., Chey, W.D., Koch, K.L., McCallum, R.W., Wo, J.M., Kuo, B., Sitrin, M.D., Katz, L.A., Hwang, J., Semler, J.R., Parkman, H.P., 2008. Differences in intragastric pH in diabetic vs. idiopathic gastroparesis: relation to degree of gastric retention. American Journal of Physiology: Gastrointestinal and Liver Physiology 294, GI384–G1391.

Hermanson, G., Silvertsson, R., 1996. Gender-related differences in gastric emptying rate of solid meals. Digestive Disease and Sciences 41, 1994–1998.

Holt, S., Heading, R.C., Taylor, T.V., Forrest, J.A., Tothill, P., 1986. Is gastric emptying abnormal in duodenal ulcer? Digestive Diseases and Sciences 31, 685–692.

Horner, K.W., Byrne, N.M., Cleghorn, G.T., King, N.A., 2014. Reproducibility of gastric emptying in overweight and obese males. Clinical Nutrition 33, 684–688.

Horowitz, M., Dent, J., Fraser, R., Sun, W., Hebbard, G., 1994. Role and integration of mechanisms controlling gastric emptying. Digestive Diseases and Science 39, 7S–13S.

Horowitz, M., Wishart, J.M., Jones, K.L., Hebbard, G., 1996. Gastric emptying in diabetes: an overview. Diabetic Medicine 13, S16–S22.

Hunt, J.N., Knox, M.T., 1968. Control of gastric emptying. American Journal of Digestive Diseases 13, 372–375.

Knight, L.C., Parkman, H.P., Brown, K.L., Miller, M.A., Trate, D.M., Maurer, A.H., Fisher, R.S., 1997. Delayed gastric emptying and decreased antral contractility in normal premenopausal women compared with men. American Journal of Gastroenterology 92, 968–975.

Lipscomb, B.L., Diggs, G.M., 1998. The use of animal-dispersed seeds and fruits in forensic botany. SIDA 18, 335–346.

Magee, D.F., Dalley II, A.F., 1986. Digestion and the Structure and Function of the Gut. S. Karger, Basel.

Michalsky, D., Dvorak, P., Belacel, J., Kasalicky, M., 2013. Radical resection of the pyloric antrum and its effect on gastric emptying after sleeve gastrectomy. Obese Surgery 23, 567–573.

Miller, R.B., 1994. Identification of wood fragments in trace evidence. In: Proceedings of the International Symposium on the Forensic Aspects of Trace Evidence. U.S. Department of Justice Federal Bureau of Investigation, Quantico, VA, pp. 91–111.

Moore, J.G., Datz, F.L., Christian, P.E., 1990. Exercise increases solid meal gastric emptying rates in men. Digestive Diseases and Sciences 35, 428–432.

Noakes, T.D., Rehrer, N.J., Maughan, R.J., 1991. The importance of volume in regulating gastric emptying. Medicine and Science in Sports and Exercise 23, 307–313.

Norris, D.O., Bock, J.H., 2000. Use of fecal material to associate a suspect with a crime scene: report of two cases. Journal of Forensic Science 45, 178–181.

Norris, D.O., Bock, J.H., 2001. Method for examination of fecal material from a crime scene using plant fragments. Journal of Forensic Investigation 51, 367–377.

O'Donovan, D.E., Hausken, T., Lei, Y., Russo, A., Keogh, J., Horowitz, M., Jones, K.L., 2005. Effect of aging on transpyloric flow, gastric emptying, and intragastric distribution in healthy humans – Impact on glycemia. Digestive Diseases and Sciences 50, 671–676.

Omür, O., Erdogan, M., Ozkilie, H., Yilmaz, C., 2014. Scintigraphic methods to evaluate alterations of gastric and esophageal functions in female obesity. Molecular Imaging and Radionuclide Therapy 23, 5–11.

Piperno, D.R., 2006. Phytoliths: A Comprehensive Guide for Archeologists and Paleoecologists. Alta Mira Press.

Rehrer, N.J., Beckers, J.E., Ten Hoor, F., Saris, W.H.M., 1989. Exercise and training effects on gastric emptying of carbohydrate beverages. Medicine and Science in Sports and Exercise 21, 540–549.

Sachs, J.S., 2001. Corpse: Nature, Forensics, and the Struggle to Pinpoint Time of Death. Perseus Books Group, Cambridge, MA.

Schneck, W.M., 2004. Cereal murder in Spokane. In: Houch, M.M. (Ed.), Trace Evidence Analysis: More Cases in Mute Witnesses. Academic Press, Burlington, MA, pp. 165–190.

Schoonjans, R., Van Vlem, B., Vandamme, W., Van Vlierberghe, H., Van Heddeghem, N., Van Biesen, W., Mast, A., Sas, S., Vanholder, R., Lameire, N., De Vos, M., 2002. Gastric emptying of solids in cirrhotic and peritoneal dialysis patients: influence of peritoneal volume load. European Journal of Gastroenterology & Hepatology 14, 395–398.

Schwartz, J.G., McMahan, A., Green, G.M., Phillips, W.T., 1995. Gastric emptying in Mexican-Americans compared to non-hispanic whites. Digestive Diseases and Sciences 40, 624–630.

Simpson, B.B., Ogorzaly, M.C., 1995. Plants in Our World. McGraw Hill, New York.

Simpson, M.G., 2010. Plant Systematics, second ed. Academic Press, San Diego.

Stacher, G., Granser, G.V., Bergmann, H., Kugi, A., Stacher-Janolta, G., Hobart, J., 1991. Slow gastric emptying induced by high fat content of meal accelerated by cisapride administered rectally. Digestive Diseases and Sciences 36, 1259–1265.

Technical Working Group on Biological Evidence Preservation, 2013. The Biological Evidence Handbook: Best Practices for Evidence Handlers. National Institute of Standards and Technology, U.S. Dept. of Commerce. http://dx.doi.org/10.6028/NIST.IR.7928.

USDA Center for Wood Anatomy Research. http://www.fpl.fs.fed.us/research/centers/woodanatomy/wood_idfactsheet.php (accessed 21.12.14.).

Winton, A.L., Winton, K.B., 1935. The Structure and Composition of Foods. Vol. 2. Vegetables, Legumes, Fruits. John Wiley & Sons, New York.

CHAPTER

5

Cases Using Evidence from Plant Anatomy

In this chapter, we describe some actual cases involving the forensic use of plant cells. Included are examples using gastric, intestinal, vomit, and fecal samples as well as some anatomy of nonfood plants. The nature of these crimes varies from burglary to homicide and rape homicide. The ways to use these simple plant anatomy techniques in forensic investigations are limited only by the imaginations of investigators and prosecutors as illustrated here.

One of the frustrations of professional forensic work comes when you have done your best, but no resolution of the case occurs. For us, one such a case occurred in Illinois where a murderer was found innocent at a retrial, even though the stomach contents of his victims indicated that he must had lied about his whereabouts at the time of the murder. The second involved the deaths of two young women in Colorado ski country several years ago that were linked by manner and cause of death as well as by their stomach contents revealing they had eaten the same last meal on the day of their disappearance. The third was the murder of an innocent 6-year-old girl in Boulder, CO, for whom justice awaits.

1. FORENSIC PLANT ANATOMY INVOLVING DETERMINATION OF TIME OF DEATH

As discussed in Chapter 4, comparison of gastric contents or vomit with a homicide victim's last meal can be useful in determining time of death. Knowing time of death helps determine who the perpetrator might be.

1.1 The Boyfriend Didn't Do It

A young college graduate working in Denver in the early 1980s failed to return home one evening to relatives where she was living. Her body was found the next day. Her last known meal was at midday with her boyfriend at a well-known fast-food restaurant known for their burgers (two all meat patties, lettuce, cheese, special sauce on a sesame seed bun) and fries. However, when the Jefferson County coroner did her autopsy, he noted that the stomach contents appeared to contain vegetable materials not available at this fast-food establishment in those days. He contacted us to examine slides he had made of the stomach contents to see if we could identify this material. We identified plants that indeed could not have been obtained at that particular restaurant indicating she had consumed another meal before she died. We found fragments of red cabbage (pigments still present), kidney beans (skin still pigmented), and onions, none of which were available at that fast-food restaurant in those days. Although the boyfriend had no alibi for the afternoon, he did have one for the rest of the day, and so he was eliminated as a potential suspect since he lacked the opportunity to commit the crime. A few years later, it was established via a confession of a serial killer that she was on her way home from work when the killer met her by chance for a second time. Believing that he was a friend of her brother (since she had accidently met him previously at her brother's home), she agreed to have dinner with the killer at a restaurant that had a salad bar, explaining the contents of her stomach found at autopsy.

1.2 The Black Widow Case

On October 21, 1993, a domestic homicide occurred in Steamboat Springs, CO. The victim was Gerry Boggs. Gerry and his brother Doug operated a very successful hardware business located on Main Street. On this day, Gerry opened the store as usual and then went down the street at about 11 a.m. to the Shack restaurant where he had his usual breakfast consisting of coffee, hash browns, eggs, and toast. Since he wasn't feeling well that day, he decided to go home and rest instead of continuing to work at the store. The next morning, he did not show up to open the store. Doug was concerned about his absence and called his home, but got no answer each time he tried during the day. Later, after closing the store, Doug went to Gerry's home and discovered his body. He had been hit on the head with a shovel, burned with a stun gun, and shot three times.

The police were immediately called in. Doug and Gerry's friends pointed their fingers toward his "estranged wife," Jill Coit, as a person of interest for possibly committing the crime. It appears that Jill had been married many times before and Gerry was husband number 8. However, some time after their wedding, Gerry learned from her seventh husband that she hadn't been legally divorced at the time of her wedding to Gerry. A longtime bachelor, this was Gerry's first marriage. He obviously was upset by this information and he had the marriage annulled and asked the court to freeze her assets, which consisted of a bed-and-breakfast business he had helped her finance. Since then, Jill had been married and divorced from husband number 9 and had acquired a new boyfriend, Michael Backus. A court date to finalize the issue between Jill and Gerry was now pending and this was considered to be a possible motive for the killing because her past behavior of marrying vulnerable men for their money would come to light. Furthermore, one of Jill's early husbands was a victim of an unresolved homicide in Texas.

On the day before the discovery of Gerry's body, neighbors reported seeing two "suspicious" characters wearing what seemed to be disguises in the vicinity of Gerry's home that afternoon. Jill and her boyfriend had an alibi for the entire evening and the next day before the discovery of Gerry's body but not for the previous afternoon. Hence, it was very critical to pinpoint the time of Gerry's death to determine if they had an opportunity to commit the crime.

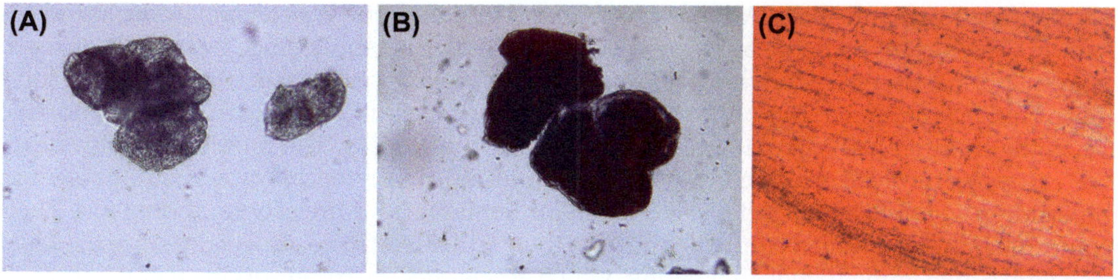

FIGURE 5.1 Potato and onion cells like those found in the stomach contents of Gerry Boggs. (A) Unstained potato cells. (B) Potato cells stained for starch. (C) Onion cells stained with safranin. *Photomicrographs by author.*

We were asked by the Colorado Bureau of Investigation (CBI) to examine the stomach contents obtained at Gerry's autopsy. The only plant materials we found were potato and onion (Figure 5.1). Since The Shack reported they did not include onions in their hash browned potatoes, it was concluded that perhaps he was killed in the evening after a second meal. Hence, Jill Coit and her boyfriend could not have been directly responsible for his death. However, we asked the investigator in the case to obtain some of The Shack's hash browns for us to examine. When the investigator observed their preparation, he noted that the cook turned the hash browns on the grill with the same spatula he used for grilling onions. Sure enough, there were some onions residing in the hash browns he was served. Thus, the analysis of his stomach contents did match his last known meal and implied that the time of Gerry's death was in the early afternoon.

Based on the stomach contents, authorities were able to obtain a search warrant for Jill's car and home. They recovered a stun gun from the car and other incriminating evidence from her residence. The stun gun was shown to produce identical marks on fresh pigskin to those found on Gerry Bogg's body. Once charged, Jill's son from a previous marriage offered up additional evidence that was used at the trial where Coit and Backus were convicted of the crime. It was after her arrest that the news media in Texas dubbed her as "the black widow" since she was suspected of the shooting death in Texas of her third husband and because she had apparently relieved other husbands of their financial assets.

The Boggs murder case led to the publication of two popular books (Singular, 1995; Linedecker, 1995), a chapter in a third book on time of death determinations (Sachs, 2001) and two television forensic shows (True TV's *Forensic Files*, "Order Up"; The History Channel's *Dead Reckoning* Disc 1 "Body Clues"), all focusing on the importance of the forensic evidence in this case.

1.3 Lizzie Borden Style Pizza

In an Illinois town, a young father, David Hendricks, claimed he left late that November night on a business trip. Earlier that evening he had dinner at 6:30 with his three children at a local pizza restaurant that catered especially to children. His wife was attending a baby shower. The children played at the restaurant's play area for over an hour after eating. According to Hendricks, he had tucked the children into their beds around 8:30 p.m. and the wife returned home at 10:30. All was peaceful when he left for Wisconsin, according to his account. The next evening he asked friends to check on his family since he could not reach them by phone. The friends were also unsuccessful in contacting them and asked the local police to check on them. The police entered the Hendricks' home and discovered a horrific site as the children had been killed with an axe while they slept. The mother was also dead. The autopsies revealed that a considerable amount of undigested pizza was still present in the children's stomachs as evidenced by tomato remnants and the aroma of oregano. Their stomachs should have been empty by the time their father left home, especially since they had been very active immediately after eating, which accelerates gastric emptying (see Chapter 4). These observations suggested that the children were already dead before he left. Taken together with other information, the husband was arrested, tried, and convicted for the killings. However, there was no direct evidence linking the father to the crime. The case was written up in considerable depth, providing a general background of the accused as well as details of the crime and trial (Vogel, 1989). However, Hendricks was later retried due to a technicality and was found not guilty. In the second trial, the gastric evidence was, unfortunately, disregarded.

1.4 Death of a Tiny Beauty Queen

On Christmas Day, 1996, the body of 6-year-old JonBenet Ramsey was discovered in her family home in Boulder, CO, sparking an intense investigation that has yet to result in an arrest for her murder. Although her stomach contained no food, intestinal contents verified that she had eaten pineapple the night before as mentioned by her parents. Fresh pineapple contains unique crystals (raphides) not found in most commonly eaten foods (Figure 5.2), making it relatively easy to distinguish. We were also asked to compare wood fragments found in tissues examined at autopsy with a possible source found at the crime scene. Later, a Grand Jury did issue an indictment, but the Boulder District Attorney chose to disregard the indictment as he believed he could not get a conviction with the evidence available at that time.

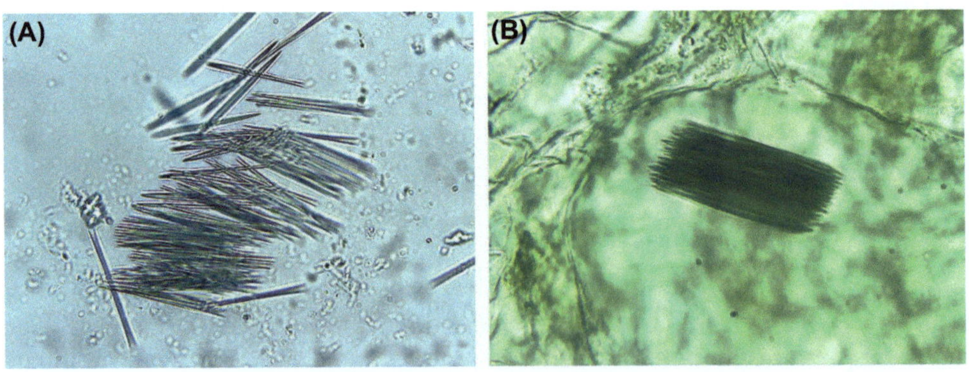

FIGURE 5.2 Pineapple raphides. (A) Loose raphides from macerated pineapple tissue. (B) Raphides packed within a pineapple parenchymal cell. *Photomicrographs by author.*

1.5 Institutional Meals Can Help Fix Time of Death

Two homicide cases from California involved victims living in confined circumstances: one a mental hospital, the other a prison for the criminally insane. Investigators were attempting to determine when the inmates died. In both cases, the victim's diet was prepared according to institutional policy. Hence, we were provided not only with stomach contents from the victims but also precise menus for all meals served prior to the death.

There were two groups of suspects in the prison. One group claimed to have seen the victim alive after lunch, but not after supper, while the second group claimed they had seen him alive after the evening meal. We were sent stomach contents of the victim and the prison dietician gave us detailed lunch and supper menus. The meals were distinct from one another. Examination of the plant materials in the stomach contents showed the victim had died after lunch, not after supper. This led the investigators to focus their efforts on the second group of inmates.

1.6 Abusive Husband Gets the Axe

In 1996, a young German immigrant killed her abusive American husband with a borrowed double-headed axe after he had allegedly attacked her while she was asleep in her bed. He allegedly had been physically and sexually abusing her since he had returned from the Gulf War in 1991. She apparently laced his beer with Nytol® and when he fell asleep, she hacked him more than 30 times with the axe while their three children were asleep. The question was raised whether this was a consequence of the nighttime attack or was it premeditated since the axe was borrowed recently by the wife. Our examination of his stomach contents revealed the presence of corn and potato fragments as well as meat fragments from the meal she had served him for supper. Furthermore, the muscle cells from the meat appeared quite fresh microscopically, suggesting the meat had been in the stomach for only a short time before his death. This finding supported the prosecutions contention that he had been killed earlier in the evening and supported the conclusion that his murder had been premeditated. However, the jury, believing the history of abuse was a mitigating circumstance, found her guilty only of second degree murder.

1.7 Sometimes Plant-Derived Food Can Be Identified

Baked goods and processed cereals usually are difficult to identify microscopically. In August 1988, a mother left her 4-year-old daughter all day (10 a.m.–5 p.m.) in her parked car while she was at work in the Denver area. That morning she had fed the child a "zinger" (similar to a crème puff; Figure 5.3). The mother claimed she checked on the child's welfare several times during the day. The last time she checked on the child at about 5 p.m., she said, she found the child dead in the car. The autopsy proved the child died from a blow to the head but it was possible this had occurred in the morning not in the late afternoon. Police thought that the morning blow was administered by either the mother or the mother's boyfriend. We examined the stomach contents and verified that the "zinger" was still recognizable in the girl's stomach when compared to a freshly purchased "zinger." This indicated she had died much earlier in the day. Faced with the autopsy information and the data on the stomach contents, the mother decided to plead guilty.

FIGURE 5.3 A zinger consisting of a flour-based covering filled with a crème.

In an unrelated case in Washington state, investigators were able to compare distinctive cereal components with the content of vomit samples associated with the victim's clothing (Schneck, 2004). Obviously, the strangled victim had regurgitated the cereal soon after eating it. A similar match was found with vomit stains present in the bed of the suspect's vehicle. It was concluded that the victim was transported in the suspect's vehicle. Furthermore, phytoliths in the soil associated with the dumpsite where the victim's body was recovered matched the phytoliths in soil recovered from the suspect's vehicle. After being arrested, the suspect committed suicide while in police custody.

2. FORENSIC PLANT ANATOMY AND AGING GRAVESITES

We are charter members of NecroSearch International (NSI), a nonprofit organization of scientists, law enforcement people, and other experts who volunteer their time and expertise to law enforcement agencies in order to locate clandestine graves (see also Chapter 9). NSI also conducts courses to train others in their crafts. They utilize a variety of sophisticated approaches to locate clandestine gravesites including aerial surveys, forward-looking infrared (FLIR), ground-penetrating radar, highly trained cadaver dogs, entomology, anthropology and archeology, forensic botany, geology, and other tools. NSI has consulted on hundreds of cases in the USA and several other countries. In the following NSI case, forensic plant anatomy played an important role.

2.1 The Cher Elder Case: Getting to the Root of the Crime

In the late March of 1993, 20-year-old Cher Elder failed to show up for her classes at Barnes Business College in Denver and also didn't appear at work. At first she was classified as a missing person, but then she was seen on a casino video in Central City accompanied by Thomas Luther, an ex-con with a history of brutal sexual assaults. According to Luther, Cher had had a fight with her boyfriend Byron Powers and Luther, who was a friend of Byron, had taken her gambling to "help her forget her troubles." The Lakewood,

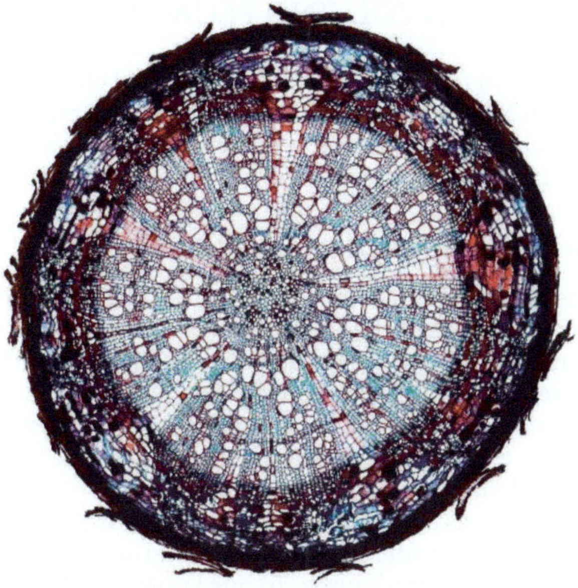

FIGURE 5.4 Cross section of a young woody eudicot root showing growth rings.

CO, police became suspicious and placed Luther under surveillance. Later that August, the police learned from an informant that Luther had bragged that he had killed a girl and buried her off of Interstate 70 where the police would never find her. Luther managed to get himself arrested for unrelated causes in April of 1995. A former cellmate of Luther's from that time told the authorities that he had been to the gravesite with Luther and he identified the general area near Berthod Pass in the mountains west of Denver. A search of the area by members of NSI with a bloodhound failed to get a positive hit on a gravesite, but they did discover a peculiar collection of rocks. The investigators took a core of soil but did not find conclusive evidence that it might be a grave. Six weeks later, Cher's former boyfriend, who was under arrest in relation to another homicide, told authorities that he also had been to the gravesite with Luther and could pinpoint the location. It turned out to be the same site already visited weeks earlier by NSI. But this time NSI investigators dug a deeper trench and NSI forensic anthropologist Dr Diane France identified the distinct smell of decomposition. NSI members carefully dismantled the site and exhumed Cher's body that was identified later by dental records. Although Luther maintained his innocence and pointed out that she could have been killed by someone and buried anytime during those 2 years, one of NSI's forensic botanists, Victoria Trammel, examined the anatomy of the roots that had penetrated the grave. Microscopic observation of cross sections of these young roots (Figure 5.4) showed that they were approximately 2-years old, indicating the approximate age of the grave. This fixed the creation of the grave to be at a time when Luther could have been responsible. Luther later was convicted of second degree homicide in 1996.

3. CASES INVOLVING FECAL MATERIAL

Comparison of plant cells in fecal samples may be useful in connecting a suspect to a crime scene. Fecal matter is often left on site following burglaries and may be matched to stains on a suspect's clothing. Similarly, rapes and rape homicides may result in the presence of fecal stains on a suspect's clothing, on upholstered furniture, or in a vehicle.

3.1 A Rape-Homicide Case Involving "Poo Prints"

A young woman was raped and murdered in 1996 after she left a party in Pueblo, CO. The suspect, who was on a work release from the local jail, met her for the first time at that party, had argued with her, and had left the party soon after she had. Later that night, he returned to the jail where the attendant noticed there were some obvious stains on the clothing he exchanged for prison garb. After her body was discovered the next day, the police investigated and identified him as a possible perpetrator of the crime. The clothing he had worn that day was confiscated as evidence.

The prosecutor in the case was aware of cases involving our analyses of stomach contents. He asked us if it were possible to identify plant material in two different fecal samples to determine if they came from the same source. To our knowledge, this had never been done[1], but we said we would examine the samples and see what we could find. We were supplied with stained pieces of cloth from the victim's clothing and the suspect's clothing as well as fecal matter from the victim. It was known that the victim, who had not been feeling well that day, had not eaten anything prior to attending the party. The last meal she had consumed was the night before, and it had consisted of Mexican style food.

Examination of the samples revealed that clothing samples from both the victim and the suspect contained evidence of black beans, chili peppers, and numerous unidentified but identical botanical parts that matched the materials found in the victim's fecal sample (Figure 5.5). Since it is unlikely that the victim and the suspect had consumed identical meals and managed to stain their clothing with fecal matter, these results strongly suggested that the suspect had been at the crime scene. This evidence was instrumental in his eventual conviction for the sexual assault and homicide.

3.2 The Church "Poor Box" Robbery

A robbery occurred at a Catholic church in Florence, CO, during which the contents of the "poor box" was taken. However, the thief left an important signature in the form of fecal material at the crime scene. Clearly the thief had experienced an uncontrolled bout of diarrhea and, although he had attempted to use a toilet, he left a considerable amount of fecal matter behind. The priest had attempted to clean up the mess before calling the police. Nevertheless, the police were able to collect some remaining fecal material from the crime scene. The police immediately suspected a local man who had a police record and was known to suffer from Crohn's disease. Sufferers of this malady often have difficulty controlling their bowels, especially if excited or anxious such as can occur during a robbery. The suspect was apprehended at the Alibi Bar, but he denied involvement and claimed he was with his sister

[1]Later we learned of an early case in New Hampshire matching human feces on a shoe to a crime scene (Johnson, 1948).

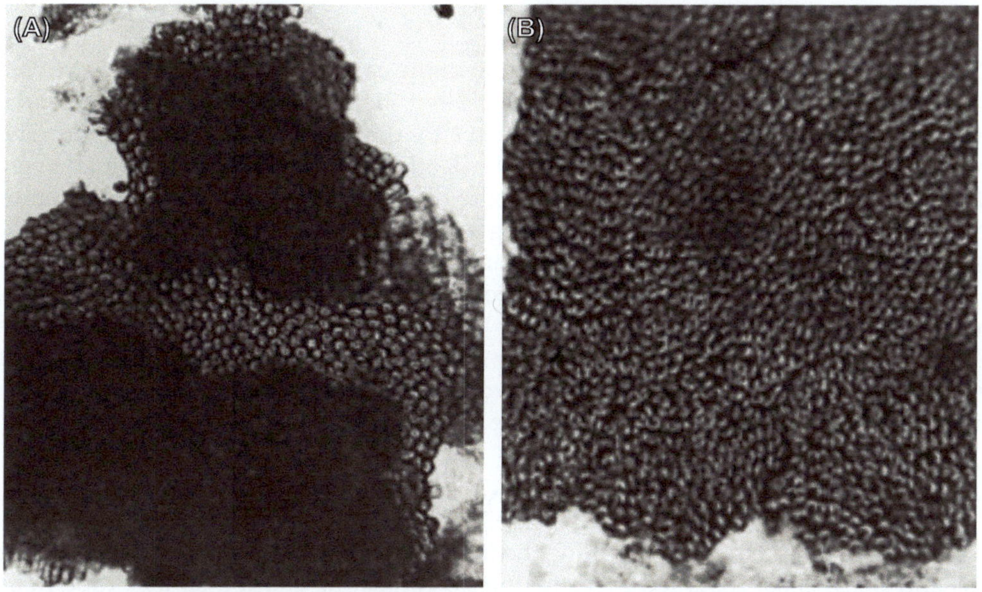

FIGURE 5.5 Match of bean skin (seed coat) from (A) suspect's clothing to (B) feces from the victim. *Reprinted with permission from Norris and Bock (2000).*

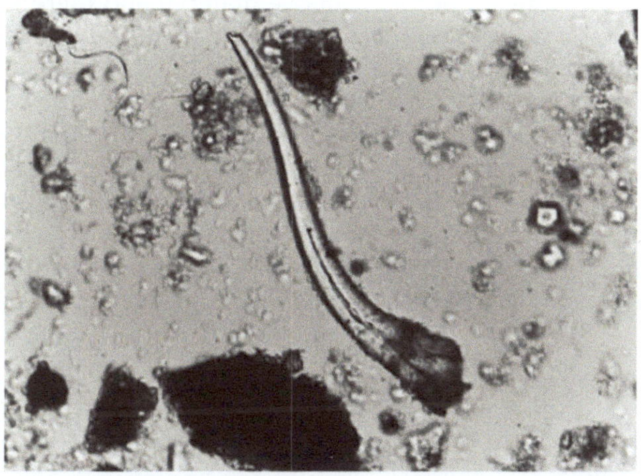

FIGURE 5.6 Example of unidentified trichrome present in a fecal sample. Even though the species of origin is unknown, the presence and frequency of occurrence of such items may be used to match two samples. *Reprinted with permission from Norris and Bock, 2000.*

at the time of the robbery. However, the police did find his blue jeans with a considerable amount of adhered fecal material in a dumpster behind the bar. When confronted with these observations linking him to the crime scene, the suspect told the police to "prove it."

We were asked if we could compare the plant cell content in the fecal material from the crime scene and the sample from the blue jeans. There were 14 items found in the two samples that matched and no items that did not match (Figure 5.6). When presented with this information, the suspect confessed to the robbery.

3.3 A Charge of Child Abuse

A 20-month-old female infant had died while in the care of a babysitter. We were asked in 1998 to examine fecal material on her diaper to see if there was evidence that she had eaten recently. The sitter's story was that she had been lethargic and had not eaten all day. The diaper was changed at about the time of the child's death. The fecal sample contained bean cells and fragments of bean seed coat as well as some distinctive but unidentified trichomes and some epidermal fragments. There was nothing available to compare to the fecal sample from the diaper. Food travels faster through an infant's digestive tract than that of an adult and it is possible that the fecal sample represented food eaten that day. However, we did not have a record of the last known meal nor of what was available to the babysitter for comparison. Apparently the investigators were looking for information that supported or questioned the story they had been told. Although we received a subpoena to testify in this case, we were never asked to report and never learned of the final outcome.

3.4 Threat of Botanical Fecal Analysis Yields a Confession

The appearance of our paper describing the forensic use of plant cells found in fecal samples (Norris and Bock, 2000, 2001) led investigators to contact us about a similar rape-homicide case in Pennsylvania. However, there were only a few small fecal stains on the suspect's clothing and because of the discovery laws governing evidence, there was great concern that an initial analysis would consume all of the available material and there would be nothing left to turn over to the defense if they were to request an opportunity to do their own analysis. Hence, we suggested that the prosecutor approach the defendant's attorney to involve the defense directly in the process by participating in the coding of the forensic samples such that we would not know their identity. They could also include some unknown samples of clothing, with and without unrelated fecal material, as internal controls. Thus, we would do the analysis blindly. We suggested that they show the defense our published papers as evidence that we could do such a comparison. As a result of our suggestions, the suspect with the advice of his attorney decided to plead guilty to third degree murder rather than take the chance of acquittal against a possible conviction for first or second degree murder if the case were to go to trial.

References

Johnson, D.J., 1948. Police science legal abstracts and notes: analysis of fecal matter as evidence of guilt. The Journal of Criminal Law, Criminology and Police Science 39, 129.

Linedecker, C.L., 1995. Poisoned Vows. St. Martin's Paperbacks, NY.

Norris, D.O., Bock, J.H., 2000. Use of fecal material to associate a suspect with a crime scene: report of two cases. Journal of Forensic Science 45, 178–181.

Norris, D.O., Bock, J.H., 2001. Method for examination of fecal material from a crime scene using plant fragments. Journal of Forensic Investigation 51, 367–377.

Sachs, J.S., 2001. Corpse: Nature, Forensics, and the Struggle to Pinpoint Time of Death. Perseus Books Group, Cambridge, MA.

Schneck, W.M., 2004. Cereal murder in Spokane. In M.M. Houck, (Ed.), Trace Evidence Analysis: More Cases in Mute Witnesses. pp. 165–190). Burlington, MA: Elsevier Academic Press.

Singular, S., 1995. Charmed to Death. Pinnacle Books, Kensington Publishing Corp., NY.

Vogel, S., 1989. Reasonable Doubt. Contemporary Books.

CHAPTER

6

Forensic Plant Taxonomy

Plant taxonomy typically involves the identification of plants to family, genus, or species based on their anatomy and/or morphology (see Chapter 3). In some cases, chemical or genetic analyses may be required, especially when dealing with identification of potentially poisonous plants or plants associated with illegal drugs. The use of anatomy/morphology can be accomplished adequately by professional botanists, experienced amateur botanists, or trained forensic scientists following procedures described here. Identification may involve plants or parts of plants associated with a suspect, a vehicle, a victim, or a primary or secondary crime scene. It may involve extrapolating from the known plants to identify a natural, agricultural, or landscape vegetation site. It may involve fieldwork at a crime scene, library, and/or an herbarium as well as laboratory or online work.

1. PLANT TAXONOMY FIELDWORK

In cases dealing with plant anatomy and the human digestive tract, we rarely are required to do anything other than laboratory work, although there was one case, the Black Widow Case (see Chapter 5), when we needed to have an investigator visit a café where the food was prepared. This was done in order to learn how more than one vegetable might occur in what the café recipe specified as consisting of potatoes alone, not potatoes plus onion. However, when we deal with investigations that depend upon accuracy in identifying plants or plant fragments using taxonomic skills, fieldwork almost always is required. It is important for the forensic botanist to visit the place or places from which the plant specimens were obtained or where investigators have questions about the plants. When extrapolating taxonomic information to a possible site, the forensic investigator should visit the site to verify that the species in question resides there.

A term tossed about with great regularity by real estate attorneys is **due diligence**. We like the term for our purposes when finding scientific names in plant taxonomy. An applicable definition for due diligence comes from www.businessdictionary.com: "Measure of prudence, responsibility, and diligence that is expected from, and ordinarily exercised by, a reasonable and prudent person under the circumstances." What follows is a version of due diligence for plant scientists dealing with forensic questions requiring answers from plant taxonomy.

1.1 Materials Needed for Taxonomic Fieldwork

The first step in doing taxonomic fieldwork is to assemble the equipment you are likely to need in some sort of field pack. An **over-the-shoulder bag** as opposed to a backpack is preferred so that you spend less time fumbling with the pack, getting it off and on. Here is a discussion of things you should keep in your field bag.

Have a **bound notebook** for recording your observations. Loose-leaf books sometimes are disallowed in evidence because of the ease of tampering with them. For example, *Rite-in-the-Rain®* field notebooks fit in your pocket and are an excellent choice as they are impervious to the weather yet easy to write in (http://www.rainwriter.com/). For note-taking, you need several **indelible ink pens or pencils** so that precipitation does not make notes unreadable.

A **hand lens** on a string around your neck, sometimes called the botanist's holy medal, is important for field observations of plants and for preliminary identifications. The hand lens should have at least 10× magnification, although 15× is ideal. In spite of Sherlock Holmes' image, magnifying glasses don't work as well. Another requirement is a **digital camera** for documenting plants at a field site. Also, a handheld **global positioning system (GPS)** unit will allow you to record the exact location so that you or others can visit the exact site at another time. Sometimes jury members may visit a crime scene along with others connected with the defense or prosecution of a case (that happened with the Mirabel case; see Chapter 9). A **local or regional flora** at hand in the field is ideal to have along, but can add much weight to your field kit. The same applies to trowels and scissors or knives for sampling. With a trowel you can collect a whole herbaceous plant if roots are needed for positive identification. Knives or scissors are useful for sampling leaves and branches from woody shrubs or trees.

1.2 Collecting Samples

Collecting bags for plant samples are necessary in fieldwork. Although botanists prefer plastic bags for nonforensic collecting, paper is preferred for our purposes whenever possible. Useful sizes range from small coin envelopes to shopping bags. In our experience, most samples will fit either into coin envelopes or paper lunch bags available at a grocery. Big paper grocery bags or paper shopping bags can be required for collecting whole herbaceous plants and smaller branches. A plant press (see Chapter 3) might prove useful in the field but is bulky. Plants collected in the field can be pressed upon returning to the laboratory.

Bring **labeling materials** as well for your collections. You should label the bags containing the collected materials with your name or initials, date, and a unique sample number that is listed in your notebook. For example, in 2015, we would assign our first case the number 15-10 and our first specimen would be 15-10-1. The next would be 15-10-2, etc. Each collection bag should be sealed securely with tape. We prefer red tape that is 2-inch-wide that you can write on with a pen or permanent marker.

On occasion, an officer of the court or another investigator may accompany you and may make separate collections. That person's collections along with yours may be preserved as evidence.

Sometimes the place where a crime was committed is not known, but the plant evidence has been collected from another source associated with the crime. For example, a vehicle associated with a crime may contain relevant plant material, or clothing found with the suspect and/or the victim may contain plants or plant parts. Sometimes these are improperly collected and/or preserved before they are given to a forensic botanist to be analyzed for identification to species. This makes her/his job very difficult. Hence, it is best to have the person doing the identification be involved in collection and processing of the plant evidence if at all possible.

1.3 Plant Identification

Laboratory work should follow fieldwork as soon as possible. Proper plant identification may require use of a simple **dissecting kit** that contains scalpels, small scissors, a ruler, forceps, and dissecting needles. These tools plus a hand lens often are all that is needed, although a dissecting microscope can be useful as well. At this time, you should press whole plants as described in Chapter 3.

To learn or confirm a plant's identification, use a regional dichotomous key or picture key. The best key is the one within your most tightly circumscribed geographic area. For example, to identify a plant found in Rocky Mountain National Park, first use Nelson's (1971) *Plants of Rocky Mountain National Park* or Weber and Wittman's (2012) field guides to the Colorado flora as opposed to using the printed volumes or online version of *Flora North America*. Some examples of floral guides are provided at the end of this chapter.

Dichotomous or branching keys are designed to walk you through the identification of a plant. If you have not used a dichotomous key, you will quickly discover that it will ask you a question related to some feature or group of features about the plant that can have only two possible answers. This is called a doublet. One answer will take you to one series of additional doublets and the other answer will take you to another series of doublets. Thus, by answering one simple question at a time, eventually you will arrive at a name for your species. Sound simple? Well it is and it isn't. It isn't quite that easy because you have to understand a lot of terms about the morphology of the plant such as is the ovary of the flower "superior" or is it "inferior." Obviously, this process requires some basic knowledge about plants and the structural features used for taxonomic purposes. We have provided some references at the end of this chapter (Harrington, 1997; Harris, 2001; Epel, 2013) that provide the basic terms necessary to use these keys. Until you have a lot of experience at this, you will need to keep these resources close at hand.

Another type of key is a color-based picture key where the authors provide you numerous pictures of plants with yellow flowers and their common and/or scientific names, then red flowers, blue flowers, etc. These types of keys usually focus on local floras only. They are often incomplete and do not list all of the species present in a region. Identifications using these color guides should be double checked by a local botanist or by an herbarium visit. Some features provided in floral guides may vary from place to place (e.g., height, flower color, etc.).

Often you may not be able to reach a firm identification on your own. If you remain uncertain of your identification, you might need to consult a local botanist or visit a college or museum herbarium. Local colleges, universities, and city or state park departments often employ botanists familiar with the local flora. Almost all herbaria require public service as part of their budgets so that these places are excellent resources for expert help. These local herbaria will have specimens collected locally for your comparison. They often have collections from other parts of the country as well as other parts of the world. Most of the world's great herbaria have now placed digital images of their holdings in public access. You can find these sources through: Wikipedia.org/wiki/Virtual_herbarium. Alternatively, you can contact an herbarium near your site to learn if they have their specimens available as digital images. Digital images are available from diverse herbaria within the US and from nations around the world. The **Kew Gardens Herbarium** in Richmond, outside of London, England, has extensive global collections as does the **Missouri Botanical Garden** (TROPICOS) in St Louis, MO.

This is also an appropriate time to make a literature search for your particular plant species. This can add to your general knowledge about the plant and teach you who is carrying out research with this species, should you need to inquire about the taxonomic details of your particular plant or plants.

Before moving on to discussing how to be sure you are using the most appropriate scientific name for a plant, a word of caution is necessary. Cultivated plants including garden plants and ornamentals can present a special problem. If you are given information that a certain plant has a limited distribution, or is especially suited to growing in a certain place such as at the crime scene, or is a cultivar whose scientific name you do not know, you can likely find the Latin equivalent from www.ars-grin.gov/, which is an index of stored germplasm for such plants. Once you get the scientific name, you should insert that name into www.plants.usda.gov/. This process provides you with the US records for where these species are common.

1.4 Authority

Verification of your identification can be the most challenging part of plant identification used for forensic purposes. There are several ways to double check your identifications. Some are fairly new, while others are over a century old. DNA analysis could be used to match pieces to the same species although in most cases not to a single plant. An example would be a tree branch piece found in an automobile belonging to someone associated with a crime. The DNA from that branch could be compared with that of a tree growing at the crime scene even if it cannot be physically matched to a particular site on the tree. However, most species examined to date do not show enough genetic variation among individuals to allow for determination of a single individual.

Chemical or biochemical analyses also might be useful for plant identification. If the plant being studied is a drug plant such as marijuana or opium poppy, there also are standard, routine biochemical tests to identify the active components of these plants, either qualitatively or quantitatively. These tests are widely used by forensic laboratories and may eliminate the use of morphological identification of the plants, for example, in marijuana cases.

Some laws concerning marijuana have listed the plant's species as *Cannabis sativa*, and plant taxonomy sometimes has been used to free a defendant. Based on minor morphological features a forensic plant *science* "expert" might claim that the sample had not come from *C. sativa*, but rather from *Cannabis indica* or *Cannabis ruderale*. Such arguments sometimes were put forward honestly, others may have been motivated by disagreement of the illegality of *Cannabis* or for financial gain. Consequently, some laws have been changed to focus on the presence of the active ingredients rather than the species of the plant. By measuring the active ingredients in the *Cannabis* and other common drug plants, one can verify that the plants are illegal in some jurisdictions. Genetic analyses are now appearing not only for the different species of *Cannabis* but even for identifying specific varieties within a species (see Chapter 3). Furthermore, analysis of pollen from other species present in a lot of *Cannabis* may identify the locale where it was grown (see Chapter 10).

The best defense of a taxonomic finding comes from a traditional approach to defending scientific plant identification. A forensic botanist working for the other side may oppose your identification of a species by accusing you of having misidentified your plant. The way to defend yourself is to turn to the scientific literature of plant science to back up your findings. This process is in accordance with the latest version of the International Code of Nomenclature for Algae, Fungi, and Plants, referred to as the "Melbourne Code", or the much older Vienna Code (see "Further Reading" at the end of this chapter).

Let's work through this topic using an example of a plant that has been used successfully in court testimony. The most used common name for this plant is water hyacinth. This semitropical species is a floating freshwater plant with beautiful blue flowers. It grows very rapidly, and often earns the designation of "noxious weed" by clogging waterways. Governmental agencies in many countries have supported efforts to control or eradicate this species from their waterways. The scientific name you have identified for this plant is *Eichhornia crassipes*, a member of the plant family Pontederiaceae. However, your opposing expert might claim that the plant is actually *Pontederia crassipes* and you have misidentified it.

If you have done your homework before going to court, you could easily counter this objection. You would have first gained access to the scientific literature for scientific plant species names by consulting the International Plant Names Index (IPNI) (http://www.ipni.org). Once you reached this site, you had clicked on the choice, *plant names*. The IPNI is a collaborative effort between the Royal Botanic Gardens, Kew, *Index Kewensis* (IK), over a million entries, *The Gray Herbarium Index* (GHI) from Harvard University of over 35,000 entries primarily from North America, and the *Australian Plant Names Index* of over 60,000 entries with greatly detailed annotations (APLI).

The IPNI provided the following: *E. crassipes* (Mart.) Solms. The "(Mart.) Solms" is the *author* for this name and identifies the person or persons who first gave it this scientifically accepted name, *E. crassipes*. This page is titled *Plant Name Query* and under "Quick Search" you inserted *E. crassipes*. The rules that govern this information are spelled out in the International Code. If you had chosen an outdated or inaccurate scientific name, this would be shown in your search, and you could then turn to the accepted scientific name.

Next, you learned more about (Mart.) Solms. The (Mart.) Solms part of the name is known as the "author" and gives us information about the people who named this species. Solms was the first person to establish this particular pairing of the genus and species for *E. crassipes*. Mart. is an accepted abbreviation of the name of the person who originally assigned the plant to the species, *crassipes*, but had placed it in another genus, *Pontederia*. Subsequent research by Solms resulted in a revision of the name to *E. crassipes*.

To see how this system catches errors, try the outdated species name for this species, *P. crassipes*. If you examine the records here, you find that this name is considered to be a base name for *E. crassipes*. This indicates it is part of *E. crassipes* history, but is not the accepted name.

By probing a little further in the IPNI site, you can learn much more about the people referred to as Mart. and Solms, but this is not necessary for documenting your assigned scientific name. For this search, three references (IK, GHI, and APNI) showed up.

They are more or less carbon copies of each other. After the name the APNI spells out *Monographiae Phanerogamarum 4: 1883*. It is here that two other botanists, whose names abbreviated as A.DC. and C.DC., first published the detailed description under the name of *E. crassipes*. As you can tell from the date, 1883, this species was written about well before DNA analyses, electron microscopes, and publish or perish policies were on hand. These monographs are written in German with extensive Latin inclusions and describe the anatomy, morphology, and other characteristics of species in great detail. The German is elegant and the details are exacting, but they are not used in courts today to our knowledge. The rules of nomenclature require use of the first name applied to a species unless later research shows that was an error.

Putting proper scientific names on plants can be challenging if you have had little experience dealing with plants, so do not hesitate to start correspondence with a major herbarium if you do not have a local one available. However, do not be a coward about trying to search on your own based only upon what you found from a local flora. In so doing, you are joining the company of scholars who have walked the same paths as you.

1.5 A Last Word of Caution

If you are trying to link two places as crime scenes or to help in the search for a likely crime scene, identifying just one species of plant from the site is unlikely to be adequate. The blame for this can be placed at the foot of our human love of gardening and horticulture. For example, just because a plant is rare and found only in one place in nature does not mean you have solved a problem. If a plant is rare, is especially attractive, or is claimed to have magical or medicinal properties, it is highly likely that the natural distribution of the species has been supplemented through cultivation. Many of the medicinal/toxic plants described in Chapter 1 are now found throughout the world. Yard and garden publications suggest in almost every issue how to make your garden more interesting with the addition of novel exotic or native plants. For example, the suggestions in *Sunset* magazine for California are based in part on research with native plants at the Rancho Santa Ana Botanic Garden in Claremont, California. The same type of service is provided by Kew Gardens near London England, and the Edinburgh Botanic Garden in Scotland is a source for ideas about rock garden plants throughout the world.

References

Aids for Plant Identification

Epel, T.J., 2013. Botany in a Day: The Patterns Method of Plant Identification. Hops Press LLC, Pony, MT.

Harrington, H.D., 1997. How to Identify Plants. Ohio University Press (Swallow Press), Columbus, OH (New edition of 1957 publication).

Harris, J.G., & Wolff Harris, M., 2001. Plant Identification Terminology: An Illustrated Glossary, second ed. Spring Lake Publications, Payson, UT.

Some General Floras and Keys to Identify Plants

Nelson, R.A., 1971. Plants of Rocky Mountain National Park. Rocky Mountain Nature Association, Estes Park, Colorado.

Weber, W.A., Wittmann, R.C., 2012. Colorado Flora: Eastern Slope, fourth ed. University Press of Colorado, Boulder.

Further Reading

Useful Online Sites

California Native Plant Society. http://cnps.org/cnps/grownative/resources.php#identification.
International Code of Nomenclature for Algae, Fungi, and Plants, http://www.iapt-taxon.org/nomen/main.php.
Plant ID Web sites. http://identifythatplant.com/plant-id-resources/plant-id-websites/.
University of Texas Herbarium, Plant Resources Center. http://www.biosci.utexas.edu/prc/.
USDA Natural Resources Conservation Service. http://plants.usda.gov/java.

Some General Floras and Keys to Identify Plants

Flora of North America Editorial Committee (Ed.), 1993. Flora of North America North of Mexico, vols. 18+ New York and Oxford.

Hallowell, A.C., Hallowell, B.G., 1981. Fern Finder: A Guide to Native Ferns of Central and Northeastern United States and Canada. Menasha Ridge Press/Wilderness Press, Birmingham, AL.

McNeill, J. (Chairman), Barrie, F.R., Buck, W.R., Demoulin, V., Greuter, W., Hawksworth, D.L., Herendeen, P.S., Knapp, S., Marhold, K., Prado, J., Prud'Homme Van Reine, W.F., Smith, G.F., Wiersema, J.H. (Members), Turland, N.J. (Secretary of the Editorial Committee), 2011. International Code of Nomenclature for algae, fungi and plants (Melbourne Code) adopted by the Eighteenth International Botanical Congress Melbourne, Australia. Koeltz Scientific Books, Koenigstein, Germany.

Tomanek, G.W., 1977. Pasture and Range Plants. Phillips Petroleum Co., Bartlesville, OK.

Some Useful Tree Guides

Harlow, W.H., 1957. Trees of the Eastern and Central United States and Canada. Dover Books, New York.

Harlow, W.H., 1959. Fruit and Twig Key to Trees and Shrubs. Dover Books, New York.

Kershner, B., 2008. National Wildlife Federation Guide to Trees of North America. Sterling, New York.

Watts, M.T., 1991. Tree Finder: A Manual for Identification of Trees by Their Leaves (Eastern US). Menasha Ridge Press/Wilderness Press, Birmingham, AL.

Watts, M.T., Watts, T., 1970. Winter Tree Finder: A Manual for Identifying Deciduous Trees in Winter (Eastern US). Menasha Ridge Press/Wilderness Press, Birmingham, AL.

Watts, T., 2004. Pacific Coast Tree Finder: A Pocket Manual for Identifying Pacific Coast Trees. Menasha Ridge Press/Wilderness Press, Birmingham, AL.

Watts, T., Watts, B., 2008. Rocky Mountain Tree Finder. Menasha Ridge Press/Wilderness Press, Birmingham, AL.

Williams, M.D., 2007. Identifying Trees: An All-season Guide to Eastern North America. Stackpole Books, Mechanicsburg, PA.

CHAPTER

7

Plant Taxonomy Cases

Taxonomy can be important forensically in a number of ways. Identification of plant fragments may link a suspect to a particular plant. Illegal drug plants must be correctly identified so that a suspect is charged correctly. Suspected toxic plants related to poisonings must also be correctly identified. The following cases illustrate some uses we have made of taxonomy in actual cases (see Chapter 9 where taxonomy also played a role in ecological analyses).

1. PLANTS CONTAINING "RECREATIONAL" DRUGS

Our general rule has been to avoid cases that deal with opiates, marijuana, and other hallucinogenic drugs. However, we made an exception for an overworked and underpaid forensic chemist we knew. He served a rural county where gangs had moved in from a nearby metropolitan area. He tended to be cynical about drug dealers, but he came across one young offender he hoped to rehabilitate. The young man had built a very large greenhouse in which he maintained a large collection of legal and illegal plants. We were asked to identify some of the plants in the greenhouse to see if they were illegal. The young man was cultivating peyote, marijuana, and opium. He'd been harvesting opium by slicing along the ripened fruits of the plants. He also was growing two species of cacti from South America. The two species were hallucinogenic, and, with a slight overdose, deadly. These cacti are often used as a base plant for grafting of ornamental cacti. These grafted cacti are readily available in stores that sell houseplants, and they are not illegal in this country but potentially are definitely dangerous.

2. "POISONED PEN" LETTERS

Mail received by certain government officials in Washington, DC, receives special attention because of their prominence. Frequently, suspicious substances are found in these officials' mail. Some suspicious substances have been forwarded to us for identification. In one case, whole seeds, split seeds, and powder were enclosed in an envelope. The samples were sent to us by the FBI to examine and possibly identify them. We did not analyze the powder chemically but focused our analysis on the identification of the seeds. With the aid of herbarium specimens, the whole and split seeds (beans) were determined to be from castor plants, *Ricinus communis* (a member of the family Euphorbiaceae). Castor beans are a source of ricin, a deadly lectin (a particular type of carbohydrate-containing protein) that is considered to be a chemical weapon. Ricin powder is extremely toxic via inhalation or injection with as little as 1.78mg being lethal to an average-sized adult. Intact castor beans are not dangerous if ingested because the seed coat is indigestible. However, the pulp from 5 to 20 beans may be fatal if ingested.

In another case, spores sent to another official were determined to be from a hallucinogenic fungus. Although it may have caused unpleasantness, it was not a lethal amount.

3. NOTHING LIKE A SHORT DRIVE AT THE GOLF COURSE

A case of malicious mischief took place in Newcastle, Wyoming, late on a rainy night. Two persons in a big pickup truck were arrested for drunk driving about a mile from a local golf course. However, the next morning the local golf course manager called the police to report someone had driven over one of their fairways and torn up a considerable amount of sod. The police still had the pickup truck in their possession. They collected dirt and sod from the tires on the drunks' truck and sent us samples from both the truck and the golf course. The grass from the truck matched the grass in the sod from the golf course.

4. "MOSS" NOT GRASS

A rape and murder had occurred in the Bahamas. The investigator in charge had taken a class we had taught on forensic botany, and so he contacted us to do an analysis. The victim's body was discovered on a golf course green. She had been raped and her throat slit. She was last seen with two men leaving a bar the night before her body was found. The two men were treated as suspects. The clothes and shoes that suspect A had worn that night were collected. However, the second man, suspect B, had disposed of his shoes and clothes. The head groundsman at this golf course planted a special strain of Bermuda grass on his greens. This grass strain was the "almond variety" of Bermuda grass and was not used on any other course on the island.

This particular Bermuda grass was extremely fine, only 2mm in diameter. The clothes and shoes of suspect A were examined with great care, and the grass found on his shoes and socks, but not on his trousers or shirt, was identified as almond Bermuda grass. The two suspects then wrote out statements accusing each other of being the rapist and murderer. The defense claimed that the absence of almond grass on suspect A's clothing meant he had not been participating in the rape and murder. Consequently, suspect A was sentenced to 10years in prison for witnessing, but not immediately reporting the murder. Mr Moss, the other suspect, was sentenced to death for the actual rape and murder. This case was documented in a televised episode of *Forensic Files* with a tongue-in-cheek title: "Moss not Grass."

5. THE BURNING TORSO

Eastern Colorado is frequently experiencing drought, and fires are an ever present danger, especially in the summer and early fall. It was no surprise to learn that a driver traveling down a rural road in County A stopped when he saw a small fire burning by the side of the road and attempted to put it out. Much to his horror, he discovered it was a burning human female torso. Her head, hands, and feet were missing. Some sections of skin had been cut out presumably to remove identifying tattoos. The body had been covered with straw to help it burn. The driver contacted the police and the fire was extinguished. Since the coroner of County A had no medical experience, the autopsy of the torso was outsourced to a private company located in a third county. Although they could not provide a positive identification, the company attempted to provide a description of who the woman had been.

Meanwhile, authorities in County B received a missing person report for a woman but she did not match the features of the torso described in the autopsy report. After several months, both cases were still open, the coroner in County B requested an exhumation of the torso so he could perform his own autopsy. The second autopsy revealed that the outsourced autopsy contained numerous errors such that the torso actually did match the approximate age and characteristics of the missing woman. The woman was known to have had tattoos in the locations where the skin had been removed postmortem.

Authorities in County B had suspected her boyfriend was responsible for her disappearance as he had a history of lawlessness. For example, he was suspected of conducting illegal cockfights, and he kept a number of chickens in his barn. We were able to match samples of the straw associated with the burning of the torso with straw from the barn, but it also matched straw from many other locations. Since chickens excrete large amounts of **uric acid** as their primary nitrogenous waste, whereas mammals excrete urea, we attempted to determine if uric acid could be detected on both straw samples. Uric acid can be detected by a simple chemical test. Straw from the barn tested positive for uric acid as expected. However, the straw from the burning site did not. We surmised that if it had been present, it was probably washed off when the fire was extinguished. We believe that this simple chemical test might have other forensic uses, and so we have included this case even though the botanical evidence was inconclusive by itself.

6. DON'T KNOW WHY...

Often when answering a request for forensic plant identifications, you may not know what inspired the question in the first place. This happens most often early in an investigation when authorities involved in a potential case are casting about for information. Here is one example of such work. An elderly man had wandered away from his residence. He was found dead on a hillside several miles away from "home." We were asked to examine the immediate area around where he was found to see if there were any poisonous plants present. The site was a grassy hillside and grasses and broad-leaved plants common to the area were present. None had any significant toxicity associated with them, although these were not considered nutritious plants for humans and were simply often consumed by indigenous wildlife species and domestic grazers. We never learned why we were asked this, but an examination of the man's stomach contents would easily have confirmed what he had or had not ingested.

CHAPTER

8

Plant Ecology

For our purposes, forensic plant ecologists usually are trained in departments that are proficient in biological or environmental sciences. The field of ecology has grown amazing speed since its beginning in the 1950s. Its broad definition, the relationship between organisms and their environments, fits many areas of human endeavor. Ecology came to the attention of the general population in 1970 when President Richard M. Nixon signed into law the **National Environmental Policy Act (NEPA)**. Within the act itself, there were several significant actions. It established the **Environmental Protection Agency (EPA)** and its many duties including the enforcement of the Clean Water Act and safeguards for our water, air, and other aspects of the human environment. The President appoints an advisory group on environmental matters called the **Council on Environmental Quality (CEQ)**. Another law signed into law by President Nixon on December 28, 1973, was the **Endangered Species Act (ESA)**. Its full name is "An act to provide for the conservation of endangered and threatened species of fish, wildlife and plants, and for other purposes." This particular act has been reviewed by the U.S. Supreme Court who found the intent of Congress was "to halt and reverse the trend towards species extinction, whatever the cost." The ESA is administered by the U.S. Fish and Wildlife Service (USFWS, Department of Interior) and the National Oceanic and Atmospheric Administration (NOAA, Department of Commerce).

These actions along with governmental employees engaged a large part of the US population. Across the country on April 22, 1970, the first **Earth Day** was held. These gatherings took many forms and Earth Day events continued to occur on each April 22 since the first one. Between President Nixon's actions and Earth Day, ecologists found themselves center stage in science, a trend that continues to the present. Academic institutions and governmental agencies demanded competent input from ecology.

Academic and commercial institutions, perhaps inspired with the impact of Earth Day, commenced to search their ranks for ecologists who seldom were identified as such—with some notable exceptions such as the Universities of Minnesota and Chicago. Demands from undergraduate and postgraduate students and young business employees inspired modifications of college curricula and hiring practices in business and governmental organizations. Ecology is now widely acknowledged as a thriving scientific field.

The field of ecology has grown with amazing speed since its beginning in the 1950s. Its broad definition, the relationship between organisms and their environments, fits many areas of human endeavor. Although originally a field within biology, claims of ecological expertise are identified by people from diverse fields such as various professional engineering, applied mathematics, sociology, and economics advocates as well as commercial spokespersons who tout their environmental products over those of their competitors. An important and dynamic subdivision of contemporary ecology is claimed by ecological modelers. They are leaders in research on climate change. They measure current global climate patterns and compare them with the past ones in order to produce predictive models of what lies ahead for earth's organisms. Of similar impact are their models of the global distribution of humans versus water and arable soils.

The forensic ecology to be discussed in this chapter involves ecological fieldwork. Unlike the evidence from plant anatomy and plant taxonomy where plant sources are the major, and sometimes only, consideration, an ecologist must have knowledge about all living things (e.g., plants, animals, fungi, bacteria) and their interactions with one another and with the nonliving or **abiotic environment** (i.e., soil, water, air, chemicals, etc.). A forensic ecologist should be a broadly trained botanist or zoologist who specializes in ecology and ecological events through time.

Ecological knowledge finds its way to court in many, many ways. The forensic ecology to be discussed here involves ecological fieldwork that usually is focused on a small geographical area. For example, this might involve linking a suspect, a vehicle, or a victim to a particular locale. Sometimes a plant ecologist (see Chapter 9) or a **palynologist** who studies pollen (see Chapter 10) is helpful in reconstructing events at a particular terrestrial site with respect to time of year. A freshwater ecologist who studies diatoms (**diatomologist**) might be useful in drowning cases (see Chapter 10).

1. AN OVERVIEW OF ECOLOGY FOR THE FORENSIC SCIENTIST

It is appropriate to discuss some common ecological terms that have scientific definitions to an ecologist, but may be confusing to others. First is the adjective **ecological**. It refers to aspects related to organisms and their environments. It does not bestow any sort of value judgment. Sometimes products are advertised as "ecological" as if they were beneficial. Ecological beauty products, cleaning substances, and foods can be seen on grocery shelves. This is an improper use of the term. An analog from chemistry is the use of organic as a value judgment. In chemistry, this simply applies to any chemical compound that contains carbon. This term includes many compounds that are deadly to humans and other organisms as well as many chemicals that make up our diets and our bodies.

To ecologists, as is true of many other disciplines within biology, the term **population** applies to a group of organisms in the same species that occupy a particular location. The term **community** is applied in plant ecology to the plants that occupy a particular place. An **ecosystem** is the combination of living and abiotic components that interact to form a more or less stable entity in a particular place.

If you are unsure of any ecological term, it can be checked in a contemporary ecological textbook, in an ecological dictionary, and the web-linked Allaby (see Further Reading at end of chapter). Constructive advice for forensic ecologists with special insights in environmental toxicology is provided by Murphy and Morrison (2014).

1.1 Ecological Landscapes

Ecologists typically become experts in relation to a certain kind of broadly defined ecosystem and often specialize in plants (**flora**) or animals (**fauna**) within that ecosystem. Ecologists identify a number of different aquatic and terrestrial **biomes**, contiguous areas that exhibit similar climates, and hence have similar inhabitants. One classification scheme identified 14 terrestrial biomes (Table 8.1) and two aquatic biomes: freshwater and marine. The 7 major biomes of North America are depicted in Figure 8.1.

Those interested in **terrestrial ecosystems** (Figure 8.2) may specialize in mountain or desert or rainforest environments, for example. Deserts can be further divided into several types including semiarid deserts and coastal deserts. Each subtype will have its own unique flora and fauna. A brief characterization of terrestrial habitat types in North America and their floral components is provided by Bryant and Jones (2006).

Even casual observers driving from the eastern plains of Colorado into the Rocky Mountains are aware of the dramatic vegetation changes that occur as the road gains in elevation. Most obvious is the relative absence of trees in the plains and that the dominant tree species observed on either side of the road in the mountains changes with increasing elevation. At the higher elevations, the trees become stunted and eventually disappear at the highest regions that still exhibit vegetation. If people were to exit their vehicles periodically, they would also note that each of these zones exhibits a unique collection of plant species (a plant community) adapted to the physical conditions present at that location (e.g., soil composition, moisture availability, temperature regimen, etc.). Similarly, each plant community coexists with a unique animal community that is adapted to the plant community and the abiotic

TABLE 8.1 The 14 Terrestrial Biomes or Major Habitat Types Identified by the World Wildlife Fund

1	Tropical and subtropical moist broadleaf forests (tropical and subtropical, humid)
2	Tropical and subtropical dry broadleaf forests (tropical and subtropical, semihumid)
3	Tropical and subtropical coniferous forests (tropical and subtropical, semihumid)
4	Temperate broadleaf and mixed forests (temperate, humid)
5	Temperate coniferous forests (temperate, humid to semihumid)
6	Boreal forests/taiga (subarctic, humid)
7	Tropical and subtropical grasslands, savannas, and shrublands (tropical and subtropical, semiarid)
8	Temperate grasslands, savannas, and shrublands (temperate, semiarid)
9	Flooded grasslands and savannas (temperate to tropical, fresh or brackish water inundated)
10	Montane grasslands and shrublands (alpine or montane climate)
11	Tundra (Arctic)
12	Mediterranean forests, woodlands, and scrub or sclerophyll forests (temperate warm, semihumid to semiarid with winter rainfall)
13	Deserts and xeric shrublands (temperate to tropical, arid)
14	Mangrove (subtropical and tropical, saltwater inundated)

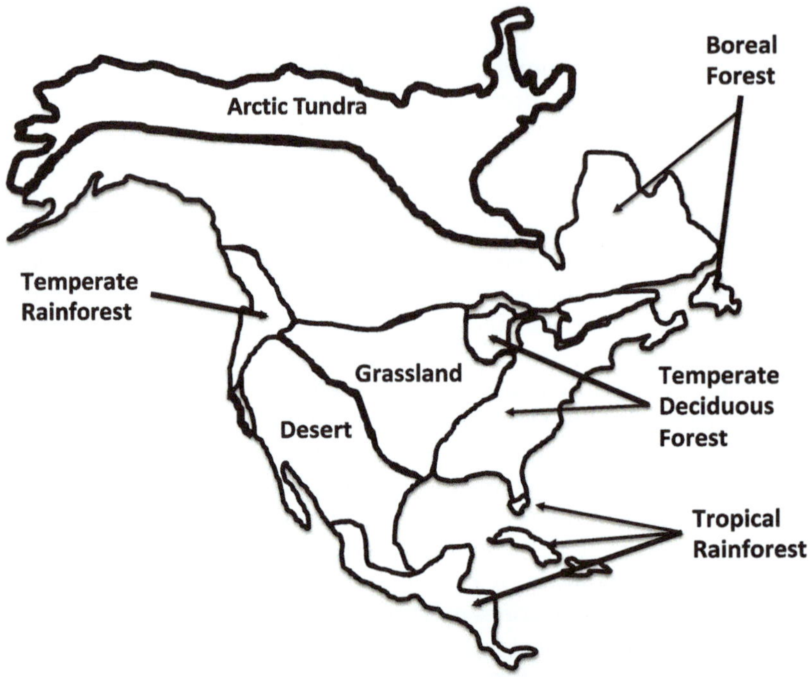

FIGURE 8.1 Terrestrial biomes of North America.

FIGURE 8.2 Some terrestrial ecosystems. Each of these can be subdivided into microhabitats. (A) Desert. (B) Rocky Mountain foothills. (C) Tropical rainforest. *Photographs by author.*

environment. Although they might see birds in each zone, there will be some different species in each zone adapted to the specific insects and/or plants present there that they feed upon. Through closer observation, they might also notice that the south-facing slopes of the mountains tend to be much drier than the north-facing slopes. Consequently, these more subtle differences within a zone result in somewhat different local communities based on the tolerance of the plants to desiccation. Although such differences may seem trivial to the casual observer, they are extremely important to ecologists and have important implications for forensic science as we describe in Chapter 9 (pp 124–126).

Aquatic ecosystems (Figure 8.3) also can be separated into major categories, each of which is studied by different kinds of aquatic ecologists. These broad categories are **freshwater** (including streams, rivers, ponds, and lakes) and **marine**. Limnologists (*limnos*, lake), for example, specialize in freshwater environments. One limnologist may be interested only in the organisms living in the bottom sediments (**benthic zone**) of a lake, whereas another might specialize in diatoms living free in the open water of lakes (**pelagic zone**). A third limnologist may study only the organisms found along the shoreline of a lake or in a pond (**littoral zone**). Each of these zones has its own unique collection of plant and animal species adapted to those specific biotic and abiotic conditions. Similarly, a marine ecologist might specialize in the plants and/or animals of the shore and intertidal region (littoral zone), the open ocean (pelagic zone),

FIGURE 8.3 Freshwater ecosystems. (A) Littoral region of a lake. (B) Alpine lake in the Snowy Range, Rocky Mountains. (C) Riparian area of a mountain stream. *Photographs by author.*

or the ocean depths (benthic zone). Still another aquatic ecologist might specialize in the unique flora and/or fauna only of **brackish water environments** that are intermediate in salinity between freshwater and seawater. A different ecologist may be interested only in **wetlands** that are intermediate or transitional between the traditional terrestrial and aquatic systems.

Obviously, each habitat type requires considerable knowledge of specific species present in the plant and animal communities, and each habitat type is studied by a special kind of ecologist. Hence, it is important to determine what kind of ecologist you may need for a certain forensic analysis before you begin your search for one.

1.2 Terrestrial Ecology

Here are some examples of how plant ecological evidence from terrestrial ecosystems may end up in court. The first example is from property boundary disputes in the USA and Great Britain. The second example of ecological forensic work involves a search for a clandestine grave.

1.2.1 Adjudication of Property Disputes through Ecological Studies

In the early US land surveys, property boundaries were recorded with such information as "the tall sugar maple tree sets the southeastern corner of the property." Here, an ecologist using an increment borer, could determine if a particular maple tree in the general vicinity is old enough to be have been present during the survey. The dimensions of tree rings sampled by the borer could tell if it had the appropriate rings to have been a "tall sugar maple" at the time of the survey. Hedgerow locations in England and part of Wales were recorded in the Doomsday Book, based on a command in 1086 by King William the Conqueror. These hedgerows set rural property boundaries for the King, the Church, and others, and had legal significance in boundary disputes. Today ecological analyses answer many questions about the early Great Britain for historians. Ecologists also have mined these data for answers to certain ecological questions about land use patterns through time.

1.2.2 Locating a Clandestine Grave Using Ecological Tools

Depending upon the lapsed time since a burial took place, there can be strong vegetation clues as to where a body is buried. Several approaches may be useful depending on the time elapsed since the victim was known to be alive.

As every gardener and farmer knows, disturbed or tilled soil is an open invitation for "weeds" to move in. Plant ecologists call these weedy species **colonizers** as they quickly establish themselves in areas following natural disasters such as winds, flood, and fires or in areas disturbed by human activities. Furthermore, ecologists speak of **succession** whereby over time, the species composition of a disturbed site may change predictably. For example, the colonizing species eventually will alter the abiotic features of the site, making it easier for other species to move in and even outcompete the colonizers. The result is a predictable change over time in the species composition of the site ultimately leading to a relatively stable community of plants called the **climax community**. The attendant animal community also changes as the plant community changes. Colonization and succession occurring after forest fires and other ecological disasters are well known to ecologists. Digging of a clandestine grave is an ecological mini disaster that necessarily disturbs the soil, no matter how careful the gravedigger is at leaving no trace. The disturbed gravesite may attract species that are different from the surrounding area, and such sites are discernible to the trained eye for months and possibly years afterward, depending on the habitat. When a body is buried shallowly, scavengers may further disturb the site and may even carry away parts of the victim. Ecologists familiar with the behavior of local scavengers may be helpful in locating missing parts in such cases.

The late Erma Bombeck (1976) wrote a book entitled *The Grass Is Always Greener over the Septic Tank*, a phenomenon that relies on these same principles. In the case of recently buried domestic animals or humans in some environments, enhanced growth and greenness of the vegetation above the site may be evident once decay has proceeded sufficiently to enrich the soil. Recently buried decaying organic matter supplies fertilizer to the plants growing above, and gives them a distinct appearance in comparison with surrounding vegetation. Depending upon the environmental conditions of a burial site, these sorts of clues can persist for as short a time as a few weeks in tropical and semitropical damp environments, but they can remain in place much longer in cold and dry habitats (Bock, 2013).

Another ground level indicator occurs in older graves as decay proceeds. Subsidence of the topsoil takes place due to the reduction in overall volume of the remains, creating a visible depression (Figure 8.4). In recent years, this has helped in the search for clandestine graves (see Chapter 9) as well as for mass graves of war victims (Brown, 2006).

Two botanists asked to search for a potential human gravesite noted that bright green vegetation covered a subsided, oblong area. Following excavation, a dog's remains were found in the grave, but no human remains were present. Although this supports the "greening effect" of a decaying body, the unnecessary excavation of the grave could have been avoided by searching with the combination of a plant ecologist and a properly trained cadaver dog. The plant ecologist recognizes the disturbed sites and the dog "hits" on a place covering human remains. If there is no "hit," the site should be marked and recorded, but not excavated at that time. Well-trained cadaver dogs, often bloodhounds, Labrador retrievers, or sometimes Alsatians, and their competent handlers make excellent collaborators.

In some other cases, questions of plant phenology become important. **Phenology** is the time line of plant development. For example, what time of year does a certain species of tree

FIGURE 8.4 Gravesite showing subsidence. *Courtesy of Jim Reed.*

produce pollen or shed its leaves. For example, the knowledge of leaf fall can be significant whether or not a crime scene is covered with leaves since either condition points the way toward when the crime could have occurred. Information on the forensic use of pollen is described in Chapter 10.

1.3 Aquatic Environments

Limnologists and oceanographers with knowledge of plant science may be needed to apply their knowledge to forensic investigations. Questions of plant taxonomy and ecology often are important here. For freshwater, brackish, and marine environments, knowledge of green algae (Chlorophyta) and diatoms (see Chapter 10) is especially important. For freshwater habitats, a working knowledge of both aquatic and semiaquatic flowering plants can be useful. Also, an awareness of toxic forms of blue-green algae (Cyanophyta) is imperative because these photosynthetic bacteria can be highly toxic to humans and aquatic organisms as well.

In marine environments, some green algae occur, but brown algae (Phaeophyta) and red algae (Rhodophyta) are more common and may need to be identified. There are only a handful of marine flowering plants restricted to the littoral zone, so they are identified rather easily.

2. PROCEDURES AND RESOURCES USEFUL FOR FORENSIC PLANT ECOLOGY

This is where an **ecological sense of place** and the matters being investigated are essential. Seasoned ecologists develop this "sense of place" through years of experience. There are a few pieces of equipment that always will be needed for fieldwork as described in Chapter 6. This includes a small portable GPS unit, a digital camera, and a bound field notebook or small recorder for note taking. A negative aspect of the recorder is that it is still essential to transcribe your notes immediately after the fieldwork or they possibly will become unintelligible after the passage of time. The benefit of recording your thoughts on site is that you do not need to pause to write on scene. In addition to these few essential pieces of equipment, the requirements for each investigation tend to call for specialized equipment.

2.1 Collection, Preservation, and Identification of Ecological Evidence

Collection and preservation of plant materials is relatively simple and generally follows logical guidelines (see Chapter 6). Care should be taken to keep specimens intact and separated by location and species, if possible. Identification can be made using published guides to the relevant local flora or by consulting local botanists. It is always wise to have the identification by amateurs verified by a qualified botanist (see Chapter 6).

2.1.1 Preservation of Terrestrial Plants

Fresh plants or large plant fragments (whole leaves, stems, roots, flowers) should be placed in paper bags (Chapter 6) and sealed or they can be prepared with the aid of a plant press as described in Chapter 3. Plastic bags will encourage growth of molds, especially in

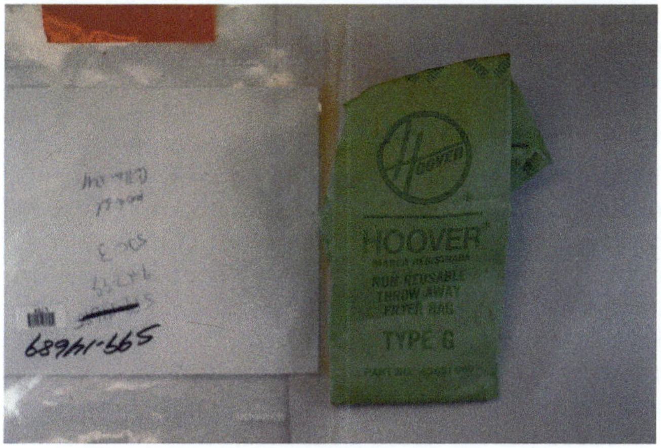

FIGURE 8.5 The wrong way to collect plant fragments.

humid climates. Small plant fragments attached to clothing, carpeting, or in or on vehicles, etc. should be carefully collected with forceps (tweezers), placed in paper envelopes, sealed, and labeled. The use of a vacuum cleaner to collect small fragments is *strongly discouraged* as it breaks up the plant fragments into even smaller pieces, making identification more problematic (Figure 8.5). Vacuum cleaners also pick up all kinds of debris that must be separated from the plant fragments. This is both time-consuming and unpleasant for the person who does the analysis.

2.1.2 *Preservation of Algae*

Marine brown or red algae can be preserved in Lugol's solution or formalin (see Appendix IV). Green algae, however, should be preserved in formalin or gluteraldehyde but not in Lugol's solution because it contains a large amount of iodine that tends to distort the cells (Graham, 1976).

Marine algae, both reds and browns make especially beautiful herbarium specimens. You put the algae in a tray of seawater or preservative, slip a herbarium sheet under the plants and position your specimen from the water on the paper. These plants produce a natural glue and will stick to the paper. An acceptable hobby for Victorian-age ladies was to make especially attractive herbarium sheets of marine algae. You can view these in older herbaria.

2.2 Ecological Resources

Here are some resources that may be useful if you plan to do your own analyses/identifications.

2.2.1 *Herbaria*

Professional botanical gardens, some museums, and many universities maintain large collections of dried plant specimens. A university herbarium typically stores collections of the local native flora that can be especially useful. Herbarium specimens are pressed flat and dried and can last for centuries. These are useful in verifying the preliminary identification of your specimen obtained from use of a plant identification key or local plant guide (see the Further Reading section at the end of this chapter).

2.2.2 Climate and Weather Data Sources

This is almost always needed in fieldwork, and you can get it before your fieldwork or afterward. Here are illustrations of what sorts of information are needed for forensic work. Suppose you are investigating a case of arson resulting in the death of a camper. It can be important that you obtain recent precipitation data in relation to the time of fire ignition or the condition of the corpse. There are several sources for such data. Two such sources are the National Weather Service for the US (www.weather.gov) and for a wider service, the National Climatic Data Center (NCDC) (www.ncdc.noaa.gov). Another service that links into those two is Weather Source (www.weathersource.com). All three charge fees, but you can avoid some or all of the costs if you have access to a business, school, or university who may be able to provide you free access to these databases. Another helpful place is Weather Underground (www.wunderground.com), although here you eventually can end up in a place that charges for the service. Another weather source greatly favored by fishermen and wind surfers is the Wind Guru that also shows tides and many other faces of weather on a daily basis around the world. It comes in a free version and a professional one (http://www.windguru.cz/int/sitemap.php).

2.3 Chemical Analyses

There are many times when chemical analyses can be very useful in forensic work, and this is true also in forensic plant ecology. For example, traces of heavy metals including zinc, lead, and cadmium along with certain other heavy metals were found on a victim's clothes. Investigators believed the body had previously been immersed in one of two lakes. The alga, *Spirogyra* and the flowering plant, *Eichhornia crassipes* (water hyacinth), were present in both lakes, and investigators collected samples of these plants from each lake and had them analyzed for heavy metal content. The plants from only one of the lakes had accumulated high levels of these heavy metals (Saygideger, 2000). Had they analyzed the water directly it would have required more sophisticated and costly analyses because of the much lower concentrations of the metals in water, whereas the plants had accumulated and concentrated the metals allowing for easier identification. Alternatively, and for much less cost, identification of diatoms associated with the body by a diatomologist might have indicated which lake was involved (see Chapter 10).

Testing of low concentrations of chemicals in aquatic samples is generally costly because of the detailed procedures required and the cost of the analytical equipment. However, researchers at many universities and environmental consulting firms can do very precise testing (to microgram or even nanogram/liter concentrations) and can analyze samples you have collected for a hefty price. However, it is more cost-effective when possible to analyze aquatic plants that have already concentrated the chemicals for you.

In addition to taxonomic knowledge of the marine environment, you may need to find tidal records, ocean current patterns, and water temperature patterns. Daily seawater temperatures are available from over 7000 recording stations around the globe (www.seatemperature.org).

A warning: this is one of those computer places where you will find out what you want to know in 2 min, and then keep on looking up places for fun for the next hour.

Marine situations can require work in/along the littoral zone, away from shore in the pelagic zone, or on the ocean floor (benthic zone). Few angiosperm species are marine and grow only in the littoral zone, so taxonomic problems with these plants are not common.

Fine nets can sometimes be used to collect small organisms. Alternatively, water samples can be collected and their contents identified using a light microscope. Here preservation of plants and animals in formalin is common. In other cases, the water must be filtered to collect the organisms or placed in a centrifuge and spun very gently to concentrate unicellular organisms. For both the littoral and the pelagic regions, water chemistry and disturbance histories can be important. For work in the pelagic zone, biological work usually is accomplished from an ocean-going vessel equipped with an onboard laboratory.

3. SOURCES OF ECOLOGICAL PLANT SCIENTISTS

Plant ecologists usually are trained in departments that are proficient in biological or environmental sciences. A trademark of all ecologists is that they wish they had taken more courses in or knew more about natural environments and the roles of members of the plant and animal communities. Hence, good ecological fieldwork often calls for teamwork because it is rare that one person will have all the necessary expertise. A common forensic team may be a combination of a geologist or soil scientist with a plant ecologist with skills in taxonomy. Soil samples plus plant fragments from a vehicle or a victim or a suspect may lead to identification of a crime scene and link a victim and/or a suspect to that site (see Chapter 9). Alternatively, the analysis may eliminate some sites being investigated while calling for detailed work at another place.

The mind-set in such work is important. That is, a forensic ecologist must employ a strong sense of place. There are ecological classifications that fit here. Are you working in a short grass prairie, a northeastern urban park, or the subalpine zone of the Sierra Nevada. If you are unable to locate a professional plant ecologist with expertise in the specific locale you need, you might enlist the aid of a local naturalist who has this sort of information. Many citizen environmentalists have obtained useful ecological and taxonomic knowledge and skills useful for local sites. The local Parks and Recreation Department may have just the person you need.

Ecologists who are competent in fieldwork find advances in their approaches to problem solving often come through cumulative knowledge rather than a "eureka moment." A founding member and the first president of the Ecological Society of America, Victor E. Shelford, serves as an example. He was born in 1877 and died in 1968. He published 141 papers and books listed in the Web of Knowledge (www.webofknowledge.com). His first paper was published in 1907 in the *Biological Bulletin.* His last publication was a 610-page book called *The Ecology of North America* published 56 years after that first paper (Shelford, 1963). In it are descriptions of all the major ecosystems in North America including plants, animals, and the physical environment of that time. Parts of the text are not accurate today because of extinctions and changes in the distribution of human populations; however, the general descriptions remain useful and accurate. Since ecological wisdom comes through years of study, perhaps there is an untapped human resource for forensic plant ecologists among the many retired academics and industrial ecologists.

Before a forensic scientist with ecological knowledge or a consulting ecologist takes on an investigation requiring ecological knowledge, it is imperative that he/she knows what questions need to be answered and whether additional expert help will be needed. For example, a case may involve fragments of several members of a plant community embedded in the organic horizon of a soil profile. If you, as a consulting forensic ecologist, do not feel capable of doing the soil work, tell the person for whom you are working and suggest where they might find one. Of course if you belong to an organization large enough that such a person is on hand, then use that person. It is best to avoid becoming a subcontractor who must negotiate the salary and the quality of the work done by the person who will be your collaborator in the work. Also, make sure you and others you work with will share all information with you. These suggestions are meant to encourage, not discourage, ecologists from working in forensic science because we have found it to be extremely rewarding and exciting work.

There are certain events where forensic plant science has not been used to our knowledge, but offer great possibilities for useful and effective forensic work. For instance, oceanic disturbances such as plane crashes and tsunamis will disturb plant distribution patterns in both the littoral and the pelagic. In plane and boat crashes, biological disturbances could be used to assist in finding the crash site and the vessel itself. This is because major disturbances to macroscopic (Yang et al., 2013) and microscopic plant life (Munari, 2013) await recolonization of the photosynthetic organisms. Aquatic colonizations often are slow processes, so search for the site of disturbance if possible. Limnologists and marine biologists would be useful additions to forensic plant ecology. Their potential in forensic plant science is little used at present.

Further Reading

General Ecology

Allaby, M., 2010. Oxford Dictionary of Ecology: Surveying the Ecological Sciences from Asteroids to Zonation. Oxford University Press, England.

Collin, P., 2001. Dictionary of Ecology and Environment, fourth ed. Peter Collin Publishing, London.

Murphy, B., Morrison, R., 2014. Introduction to Environmental Forensics, third ed. Academic Press, San Diego, California.

North America Flora Guide

Shelford, V.E., 1963. The Ecology of North America. University of Illinois Press, Urbana.

Local Flora Guide Examples (e.g., Rocky Mountain region)

Ells, J., 2011. Rocky Mountain Flora. Colorado Mountain Club Press, Golden, CO.

Weber, W.A., Wittmann, R.C., 2012. Colorado Flora: Eastern Slope, fourth ed. University Press of Colorado, Boulder.

Some Online Resources

Brown algae

http://www.nps.gov/acad/naturescience/brownalgae.htm.

Red algae

http://www.marineplantbook.com/marineplantbookredalgae.htm.

Weather
http://www.windguru.cz/int/sitemap.php.
www.weather.gov.
www.ncdc.noaa.gov.
www.wunderground.com.

Global seawater temperatures
www.seatemperature.org.

USA tide data
http://tidesandcurrents.noaa.gov/map.

References

Bock, J.H., 2013. The use of macroscopic plant remains in forensic science. In: Elias, S.A. (Ed.), The Encyclopedia of Quaternary Science, vol. 4. Elsevier, Amsterdam, pp. 542–547.

Bombeck, E., 1976. The Grass is Always Greener over the Septic Tank. McGraw-Hill (Available in paperback, Fawcett Publishers, 1995.).

Brown, A.G., 2006. The use of forensic botany and geology in war crimes investigations in NE Bosnia. Forensic Science International 163, 204–210.

Bryant, V.M., Jones, G.D., 2006. Forensic palynology: current status of a rarely used technique in the United States of America. Forensic Science International 163, 183–197.

Graham, L., 1976. http://www.botany.wisc.edu/courses/botany_330/preservation.html.

Munari, C., 2013. Benthic community and biological trait composition in respect to artificial coastal defense structures: a study in the northern Adriatic Sea. Marine Environmental Research 90, 47–54.

Saygideger, S., 2000. Sorption of cadmium and their effects on growth, protein contents, and photosynthetic pigment composition of *Veronica anagallis-aquatica* L. and *Ranunculus aquatilis* L. Bulletin of Environmental Contamination and Toxicology 65, 459–464.

Shelford, V.E., 1963. The Ecology of North America. University of Illinois Press, Urbana.

Yang, S., Wheat, E., Horwith, M., Ruesink, J., 2013. Relative impacts of natural stressors on life history traits underlying resilience of intertidal eelgrass (*Zoster marina* L.). Estuaries and Coasts 36, 1006–1013.

CHAPTER

9

Forensic Plant Ecology Cases

Individuals may be connected to a crime scene through the examination of fragments of plant material. Vehicles often carry a recent history in the fragments of plants collected inadvertently as they travel through a particular habitat. These fragments may be found inside the vehicle, underneath the body, in the engine compartment, around the wiper blades, etc. Sometimes plant fragments may be dumped with the body or embedded in the clothing of the victim or the suspect. Often, the source plants may be identified from whole or partial stems, leaves, flowers, or seeds. The presence of several species of plants may indicate a particular habitat with which a suspect, body, vehicle, etc. was previously associated. Sometimes a plant fragment associated with a suspect or an object can suggest the type of habitat where the body may have been hidden.

Questions of jurisdiction can arise concerning as to where a homicide took place because that is the likely place for an investigation to be centered and a possible trial to be held. If county boundaries are crossed, trials are the financial responsibility of the city or county where the crime occurred, not where the body was discovered. If state boundaries have been crossed, this may be a federal case. Plant fragments associated with the victim, the suspect, or a vehicle may be used to answer questions of jurisdiction. For example, plant fragments associated with a body dumped in Iowa may actually be from species unique to the eastern slope of the Colorado Rocky Mountains suggesting the homicide occurred in Colorado, not Iowa. Similarly, pollen, insects, and soil samples may provide significant clues to a particular habitat.

Knowledge of plant ecology also can be very useful in the location of clandestine graves. For example, disturbance of surface soil often allows weedy species to colonize the disturbed area. Thus, the disturbed site can be recognized by a trained observer due to its unique collection of species that differs from surrounding undisturbed habitat.

1. USES OF FORENSIC PLANT ECOLOGY CONNECTING SUSPECTS TO CRIME SCENES USING PLANT FRAGMENTS

These cases successfully linked a suspect or a suspect's vehicle to a particular habitat based upon the species of plants identified from plant fragments.

1.1 Death of an Abused Mother

In southeastern Colorado, Jacklyn Funderberg, a mother of a young infant, had her jaw broken by her boyfriend. The boyfriend then moved back in with his ex-wife. Soon afterward, Jacklyn disappeared, and her remains were found buried in a shallow grave at the base of a rocky outcrop on the eastern prairie (Figure 9.1). A distinctive shrub was growing on top of the outcrop that is unique to this kind of formation. Fragments of this shrub also were associated with the victim's body as well as grasses found only at the gravesite. Investigators surmised that she was thrown off the top of the outcrop and then buried at the bottom. The investigation focused upon the ex-boyfriend. When we examined his vehicle, it was surprisingly free of all types of debris including plant fragments; i.e., it had recently been thoroughly cleaned. While in custody, the boyfriend phoned his ex-wife and asked her to wash some of his clothes and to hide a pair of shoes. However, the Sheriff's Department used his recorded phone message to obtain a warrant and seized his clothing from his ex-wife's washing machine. We showed that the suspect's clothing was contaminated with pieces of the unique shrub growing at the top of the outcrop as well as parts from the low-elevation grasses found near the grave. Upon learning of this evidence, the suspect changed his original story and claimed during the trial that they had both planned to jump off the top of the outcrop but that he had lost his nerve when it came his turn. He then panicked and buried her body. This indirect plant evidence, suggesting he had been to the crime scene or a very similar location, was part of the trial evidence that led to a conviction of first-degree murder.

1.2 Gang-Related Abduction and Sexual Assault of a College Girl

In 1999, a college girl was kidnapped as she walked along a street one night in Boulder, CO, by six young men from the Denver Asian Crips gang. They took the girl in their van to a relatively remote area in nearby Left Hand Canyon where she was sexually assaulted. They released her in the same area and she sought help from a nearby resident. An observer of the abduction had reported the make and part of the van's license plate. The vehicle was

FIGURE 9.1 Site where body was recovered. Body was thrown from top of cliff and then buried at arrow. *Photograph by author.*

found quickly by local police, and impounded at the police garage. An officer collected plant fragments from inside the car on the front and back seats, the carpets, the car's pedals, the tire treads, and the window wiper wells. These collections were placed in individual stamp envelopes, stored in a large paper bag. We visited the place where the girl remembered being taken, and we matched the collected specimens from the vehicle to the vegetation growing at the crime site. The site was at a higher elevation from the place where the kidnapping occurred and where the assailants lived. The vegetation at this elevation differed significantly from the plants found in the Boulder–Denver plain areas. The suspects' van had definitely been at that elevation recently although it was not possible to conclude it had been at that exact site since the vegetation is the same in other canyons at that elevation. Nevertheless, this was useful circumstantial evidence used for obtaining arrest warrants, and again later in court. Five of the men involved were charged with the crime. The sixth man had committed suicide when police attempted to capture him in Michigan. The victim was able to identify each assailant by the sound of their voices. The trial resulted in conviction of the five surviving suspects (Figure 9.2). As a footnote to this case, the victim went to law school and now works in law enforcement.

THE ATTACKERS

Sonny Lee
Age: Now 28
Status: Maximum-security inmate at the Limon Correctional Facility.
Sentence: Two consecutive prison sentences of 36 years to life for sex assault, plus 36 years for kidnapping
Parole: Eligible in 2092

Johnny Lee
Age: Now 22
Status: Maximum-security inmate at the Sterling Correctional Facility.
Sentence: Thirty-six years to life for sex assault, plus 36 years for kidnapping
Parole: Eligible in 2089

Kao Vang
Age: Now 23
Status: Transferred to the California Department of Corrections in 2001
Sentence: Pleaded guilty to second-degree kidnapping and attempted second-degree sexual assault. Kao Vang, Chue Vang's cousin, had his 32-year prison sentence reduced to 11 years in exchange for testifying against Sonny Lee and Johnny Lee, who are not related.
Parole: Information not available

Johnny Lee | Sonny Lee | Chue Vang | Kao Vang | Steve Yang

Chue Vang
Age: Now 21
Status: Transferred to the super-maximum-security Colorado State Penitentiary in Cañon City after 2003 contraband violation.
Sentence: Twenty-four years to life for sex assault, plus 16 years for kidnapping
Parole: Eligible in 2023

Steve Yang
Age: Now 24
Status: Medium-security inmate at the Arkansas Valley Correctional Facility near Pueblo
Sentence: Sixteen years to life for sex assault, plus 16 years for kidnapping
Parole: Eligible in 2022

Kather Yang
Kather Yang killed himself in September 1999 as police in Green Bay, Wis., tried to arrest him. He was 20.

Source: Colorado Department of Corrections and Camera files

FIGURE 9.2 Gang responsible for the rape in Left Hand Canyon. *From Boulder Camera.*

1.3 "Her Vehicle Had Never Been in the Mountains"

In September 1999, Matthew Mirabel reported that his 24-year-old wife, Natalie, was missing after she failed to return from a late night trip to the grocery store in Longmont, CO. He had stayed home with their infant daughter while she was gone. Her car was found the next morning in the parking lot of the grocery. Later, her strangled and decapitated body was found in a nearby canyon above 7000 ft. A couple of murders involving decapitation had recently occurred in nearby Denver. Natalie also had been bludgeoned with a segment of wood broken from the end of a piece of 2×2 board that was found with the body. A comparison of the plant fragments associated with her vehicle, the dumpsite, and her Longmont residence indicated that the vehicle had recently been at an elevation in the mountains similar to the dumpsite. Her husband insisted her car had never been driven to the mountains.

Another piece of botanical evidence proved important in this case. The Sheriff's investigators discovered an end piece of a 2×2 board in the back of the husband's pickup truck but the broken end was not a match to the piece found at the dumpsite. However, investigators did discover a third segment of 2×2 along the highway between the Mirabel home and the dumpsite that proved to be the missing piece connecting the broken ends from the dumpsite piece and the piece found in the husband's vehicle (much like the ladder pieces and wood from Hauptmann's attic in the famous Lindbergh case).

It was later discovered that Matthew was having an affair with his brother's wife, and that he had attempted to purchase a $1 million life insurance policy on Natalie but the insurance company would only sell him one for $250,000. Furthermore, a glove discovered in Natalie's car had both Natalie's and Mathew's blood on it. He apparently had driven her vehicle to the mountains to dump her body. Matthew was found guilty by a jury of his peers in June 2000.

2. USES OF FORENSIC PLANT ECOLOGY IN THE LOCATION OF CLANDESTINE GRAVES

Understanding the ecology of a particular region can be very helpful in the location of clandestine graves. As described in Chapter 5, we are charter members of NecroSearch International (NSI), a nonprofit organization of scientists, law enforcement people, and other experts who volunteer their time and expertise to law enforcement agencies to aid in the location of clandestine graves. In the following NSI case, forensic plant ecology played an important role.

2.1 The Michelle Wallace Case: A Spruce Needle Was the Key to Location

Twenty-five-year-old Michelle Wallace was on a photographic hiking trip in the central Colorado Mountains with her dog, Okee, in the summer of 1974. When she failed to check in with her parents back in Illinois, they alerted Colorado authorities to locate her. Extensive searches were conducted near Crested Butte where she was last believed to have been hiking, but no trace of Michelle was found. Later a local rancher shot and killed Okee when he found the dog harassing some sheep. This case has been covered in several publications (Gordon, 2007; Wyman, 2010) and TV programs (e.g., *Cold Case Files*, A&E).

Although no trace of Michelle could be found after numerous searches of an expansive area of wilderness, her car was discovered in Amarillo, TX. Authorities then received a tip from Chuck Matthews, who had last seen Michelle in the company of Roy Melanson on the day she disappeared. At that time, Melanson was wanted for skipping out on a previous rape charge in Texas. When Melanson was apprehended for the previous crime by police in Pueblo, CO, he had in his possession Michelle's wallet as well as pawn tickets for some of her belongings. When questioned, he claimed that he had stolen her car and belongings but to his knowledge she was alive when he last saw her. Police recovered her belongings, and found that the last picture on an undeveloped roll of film still in her camera was a picture of Melanson. Nevertheless, there was no definitive evidence to prove Melanson had murdered Michelle, and he was extradited to Texas where he was wanted on the rape charge.

In July 1979, some hikers on a remote logging road near Kebler Pass, an area somewhat distant to the area searched previously for Michelle, came upon a scalp with long braids similar to those Michelle had when she disappeared. Although this specimen was turned over to the local Sheriff, nothing further happened until 1991 when Chief Investigator, Kathy Young, of the Gunnison Police Department decided to go through the collected evidence from the now cold Wallace Case. Adjacent to Michelle's effects, she discovered an unlabeled cardboard box in the Gunnison Sherriff's Department containing the scalp and braids. Young sent samples of hair from Michelle's hairbrush along with hair from the braids to the Colorado Bureau of Investigation who declared them to be a match. She then asked NSI to attempt to locate the body/gravesite near Kebler Pass. Victoria Trammel, one of the NSI's forensic botanists, examined the braided hair and noticed it appeared to be sun-bleached, indicating that the body had probably not been buried. She found numerous subalpine fir needles (a species frequenting wetter slopes) embedded in the braids, however, there was only one Engelmann spruce needle among them (a species that likes dry, sunny slopes), suggesting that the body might be found on a more moist slope. In 1992, NSI volunteers examining a moist slope in the vicinity of where the scalp had been found discovered an exposed skull containing teeth, one of which sported a gold crown that matched Michelle's dental record. Additionally, they located a hiking boot containing foot bones as well as other bones nearby (e.g., femurs, vertebrae). These discoveries were in the habitat predicted by the ecological botanical evidence.

Meanwhile, investigators had kept track of Melanson over the years since they were convinced he was responsible for Michelle's disappearance. In the interim since Michelle's disappearance, Melanson had been convicted of rape in Texas, served 14 years, and was released in 1988. At the time the body was discovered, Melanson was incarcerated in Kentucky for robbery. On the anniversary of Michelle's birthday, Kathy Young served Melanson a warrant for his arrest. He was tried in Colorado and the jury deliberated only 5 h before convicting him of the murder of Michelle Wallace. Her remains were laid to rest next to her mother who had committed suicide after Michelle's disappearance, providing some closure after 19 years for her surviving father, George Wallace, and a brother. Sadly, George, at age 85, was killed in his Florida home by intruders in 2006.

The discovery of Michelle Wallace's body paid off with some additional dividends. While serving his sentence for the Wallace murder, Melanson's DNA and fingerprints were connected to the murder of 51-year-old Anita Elizabeth Andrews that had occurred in Napa, CA, just a few weeks before Michelle's murder. Melanson was convicted of that crime as well in

2011 (see Shulman, 2012). DNA also connected him with the 1988 murder of a 24-year-old Louisiana girl, Charlotte Lily Sauerwin, that occurred soon after Melanson was released from the jail in Texas.

3. OTHER USES FOR FORENSIC PLANT ECOLOGY

3.1 Uprooted Plants Can Tell Time

Early in our Colorado forensic work, both forensic botany and forensic entomology were still viewed with some skepticism by criminalists. One summer, a corpse was found along a dirt road in the eastern Colorado plains. Sunflowers growing nearby had been uprooted and used to cover the body. The plants were dried and wilted but were still green in color. Those sunflowers were provided to us and we froze them to stop further changes. Fresh sunflower plants from the same site were brought to the lab. We were asked to determine how many days had passed since the sunflowers were pulled from the ground to form a makeshift shroud. We put our sunflowers in a rooftop greenhouse where we simulated the conditions on the eastern plains as best we could. Their wilting was checked daily for 3 weeks. We determined from the wilting of the greenhouse plants that the plants covering the victim must have been in place between 7 and 11 days. Wilted sunflower plants from earlier and later time periods did not match those covering the corpse. Unknown to us, a forensic entomologist was given the same question using his specialty: How long had the insects on the corpse been in place? He came up with an estimate of a week or a couple days more. The results cast suspicion on an alibi provided by the principal suspect in this case because his alibi did not cover the time period estimated by the two different approaches. Thus, another simple application of forensic plant science was established.

3.2 Phenological Indications: Plant Patterns in Central Florida

This case was a death penalty case. A little girl disappeared in mid-summer in Orange County, Florida. Her skeletal remains were found in December of that same year. A forensic botanist reported that the remains likely remained where they were found in December because leaf fall occurred in early fall and the site was covered with a significant level of leaf fall. It is schoolbook knowledge that trees lose their leaves in fall even though that is untrue for many places. An Inuit friend told me that was a question she missed about trees on an IQ test once because she'd never seen a tree. Also, leaf shed can be governed by wet versus dry seasons. In southeastern Arizona, live oak trees shed their leaves in response to drought while in coastal California, trees of the same species drop them all year round.

The pictures taken at the place where the remains were found showed a heavy accumulation of leaves over the site. After the discovery, the site was scraped clean of all debris because the investigators wanted to make sure they did not miss any bones or other clues. In fact, the FBI screened the debris with great care. When the site was first seen in person by Bock a couple of weeks after the discovery, the area was swept clean. After examining the site and studying the list of plants at the site, it was obvious that there was a significant semitropical element to this flora. And when the site was revisited near the end of February, the area once more was covered by a heavy leaf litter. This leaf cover compared favorably to that from the photos taken the day the remains were discovered.

A second aspect of the same case called for ecological knowledge pertaining to one of the larger skeletal bones. It was found buried under approximately 15 cm of debris and soil. The area is near a school and some housing developments. It is likely that the area was visited by dogs. Dogs are well known for burying bones. Also, coyotes sometimes bury bones (M. Bekoff, personal communication). Coyotes in considerable numbers had been spotted in this region in recent times so perhaps one or more bones of the human skeleton might have been cached by a canid.

3.3 A Case in Progress: Open Investigation

The next case also involves the death of a child whose body was found inside her home. Certain of the detailed evidence we worked with cannot be discussed even more than a decade after the crime. The case remains an open homicide investigation for which there is no time limit. Part of our findings was reported to a Grand Jury, who prepared an indictment but the District Attorney refused to activate it. One of the scenarios under investigation was that the murderer had entered the house via a certain small basement window that had a broken pane. This scenario is possible, except for one plant clue. The soil beneath this small window was covered with healthy Christmas rose plants (*Helleborus niger*: Family Ranunculaceae). These are thin leaved, green plants with pink flowers that tolerate dank cold weather. They bloom around Christmas time and were robust at the time of the homicide. They showed no signs of disturbance, no crushed leaves, no broken petioles. This means the window likely was not used to enter the house because for even a small person it would have required considerable struggle. Some other possibilities are an outsider might have come in by some other entrance, or perhaps the murderer(s) were residents of the house. Another botanical puzzle found on the corpse of the young victim was a piece of green moss. We were not allowed access to the premises to survey what mosses were bright green at the time of the crime. Those Christmas roses and mosses continue to haunt us.

References

Gordon, O., 2007. Cold Case File: Murder in the Mountains. Bearport Publishing.

Shulman, T., 2012. Murder and Mayhem in the Napa Valley. The History Press.

Wyman, A., March 22, 2010. Two Women May Share a Common Killer. Napa Valley Register. http://napavalleyregister.com/news/local/two-women-may-share-a-common-killer/article_6ab4e342-34a5-11df-9281-001cc4c002e0.html.

CHAPTER

10

Additional Approaches in Forensic Plant Science

Several aspects of plant science in addition to plant anatomy, plant taxonomy, and plant ecology of seed plants have been employed forensically although we have done little work in these other areas. Genetics was briefly discussed in Chapter 3. In this chapter, we focus on forensic uses of pollen and algae, particularly diatoms. Both of these topics relate to forensic plant ecology and are closely tied to plant taxonomy. Other plant groups, including fungi, bryophytes (e.g., mosses), and ferns, also may have forensic applications although they have not been utilized very often.

1. PALYNOLOGY

Palynology is the study of pollen grains and some related structures such as fern and fungal spores. Together, these forms are termed **palynomorphs**. This area was pioneered at the beginning of the twentieth century by a Swedish geologist, Lennart von Post. We are including this plant science topic as it has important implications especially for ecological aspects of forensic plant science. Detailed discussions of the field of palynology are readily available (e.g., Milne et al., 2005; Bhattacharya, 2011) and only an overview is provided here.

Fossil pollen has been used successfully by geologists to characterize oil and gas deposits and hence to use these data to locate new sources. Art appraisers have used pollen to authenticate old paintings. Climatologists use fossil pollen to reconstruct past climates. Anthropologists can deduce diets of ancient peoples from pollen grains associated with tools used for food preparation or from pollen grains present in preserved feces (coprolites). Pollen grains also have a wide range of uses in both criminal and civil processes (Table 10.1).

Although pollen analyses occasionally have been used in court cases, forensic palynology has been used sparingly in the US, and there are very few palynologists with any forensic experience here. New Zealand is perhaps the world leader in forensic palynology today. One US center for forensic pollen work is located presently at Texas A&M University where there is a pollen reference collection for 20,000 species of plants (Bryant and Jones, 2006).

TABLE 10.1 Forensic Uses of Pollen Grains[a]

1.	Relate a suspect to the scene of a crime or discovery scene
2.	Relate an item left at the crime scene or discovery scene to a suspect
3.	Relate an item at the discovery scene to the crime scene
4.	Prove or disprove alibis of suspects
5.	Narrow down a list of suspects
6.	Determine the travel history of items, including food and drugs
7.	Provide information as to the environment that an item has come from
8.	Provide information as to the geographic source of items
9.	Determine season of year when an event occurred
10.	Aid police in their lines of inquiry
11.	Help locate clandestine graves and human remains
12.	Help determine the perimortem fate of a victim
13.	Help to determine the deposition period of human remains

[a] *Modified with permission from Mildenhall et al. (2006).*

Pollen has several features making it very attractive for forensic studies (Table 10.2). Moreover, the great variety of described plant species and their known distributions in North America suggests that palynology is essentially an untapped resource that begs for more utilization by forensic scientists.

TABLE 10.2 Characteristics of Pollen Making It Useful for Forensic Purposes

1.	Pollen grains are so small as to be invisible to the naked eye and hence easily overlooked by criminals.
2.	Pollen grains are extremely resistant to degradation.
3.	Each species of seed plant produces a structurally unique pollen grain.
4.	Specific habitats can be identified by the pollen grains of their plant inhabitants.
5.	Pollen patterns vary predictably throughout the year for each location.

1.1 Biological Features of Pollen

Pollen grains represent the sperm of the plant world and are produced by meiosis hence pollen grains are haploid; (see Chapter 3) in angiosperms by the **stamens** of male flowers and by the stamens of complete flowers (flowers having both male and female parts). The female part of a flower consists of one or more **pistils**. A pistil has a **stigma** at its tip where pollen may attach, and a stalk or **style** that continues to the **ovary** at the base. Male cones of conifers also produce pollen and the ovaries are located on female cones. Each species of the more than 300,000 described species of flowering plants and conifers produces a structurally unique type of pollen grain. Thus, **pollen prints** can be useful for identification of a single species or of a region consisting of a particular grouping of plant species.

Each pollen grain has an outer impervious cellulose coat, the **exine**, that is strengthened with **sporopollenin** and other chemicals secreted by the parent plant. The result is a decay-resistant outer structure that can survive for millions of years after the core has decayed. Within the pollen grain there is one or two additional layers surrounding two nuclei, one of which will eventually provide the male gamete.

Pollen grains are essentially invisible to the naked eye, varying from 5 to 200 μm in diameter (there are 1000 μm in 1 mm; 200 μm = 0.2 mm). Most pollen grains are about 25–40 μm in diameter. Consequently, the casual observer routinely overlooks the presence of pollen and its potential for forensic analysis. Furthermore, each genus or species of plants produces a pollen grain with a uniquely shaped and/or sculptured exine within a characteristic size range. Thus, it is possible for a pollen specialist to identify the source microscopically to family, genus, and often to species.

Pollen grains vary greatly from species to species in colors and shapes. They may be spherical, angular, multisided, etc. (Figures 10.1 and 10.2). They may also possess various

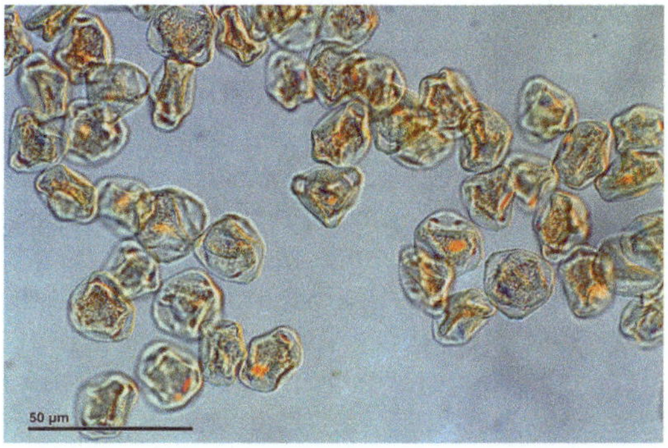

FIGURE 10.1 Light photomicrograph of pollen grains from common hazel, *Corylus avellana. Pollen grains, courtesy of Josef Reischig, available at: http://commons.wikimedia.org/wiki/File:Pollen_grains_%28251_20%29_Pollen_grains_of_common_hazel_%28Corylus_avellana%29;_total_preparation.jpg.*

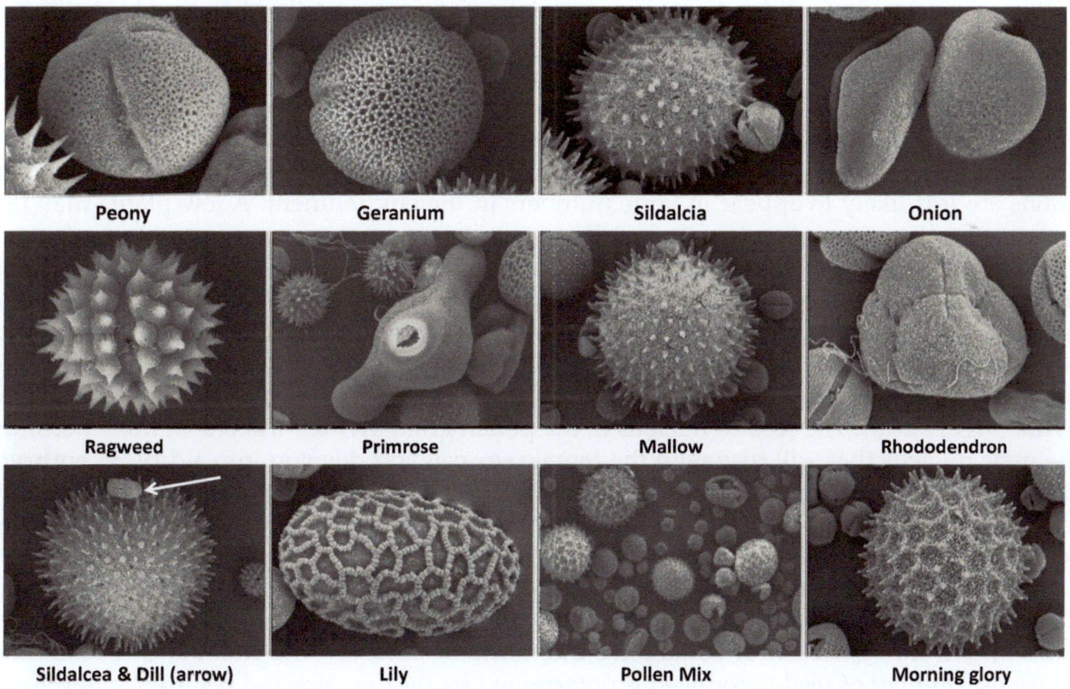

FIGURE 10.2 Scanning electron micrographs of pollen grains. *Darmouth College Electron Microscope Facility by Lisa Howard. http://remf.dartmouth.edu/pollen2/pollen_images_3/index.html.*

protrusions that can vary in shape among different species. Most pollen grains have tiny openings or apertures through which the male gametophyte (see ahead) can escape after reaching the appropriate female structure. Apertures may be as simple as a single pore or slit. Pollen grains of other species may exhibit multiple complexes of pores and slits. The surface of the pollen grain of a particular species may be plain or elaborately sculptured with depressions or ornamentations (protrusions) that can further aid in the identification of the source of the pollen.

Pollen reaches the female parts of flowers and cones (a process called **pollination**) through transport by various modes. Most flowers and cones are pollinated by actions of the wind that randomly blows pollen from its male source to the female structures. Because pollination by wind is a random event, these wind-pollinated plants produce huge amounts of pollen. For example, a single male *Cannabis* flower may produce 60–80,000 pollen grains, whereas a single complex male tassel of an alder tree that consists of many individual flowers may release 4–6 million pollen grains. A medium-sized black pine reportedly can release 10 billion pollen grains in one season. A single bracket fungus, *Ganoderma applanatum*, can produce 30 million spores per day from May to September. Pollen grains and fungal spores dispersed by winds are the basis for the misery experienced by thousands of allergy sufferers every year. On a Texas spring day, there may be 500 pollen grains and fungal spores per cubic meter of air, and an inactive person spending the day indoors will inhale 7200 pollen grains that day (Bryant, 2015). Imagine how much an active person in the outdoors might accumulate.

Some flowers rely on animals, such as a variety of insects as well as bats and birds, to perform this transfer of pollen. Furthermore, the flowers of these plants are often constructed so that only one or only a few related animals may accomplish this essential feat for the plant. The appropriate animal visits one plant searching for nectar and/or pollen and carries away pollen from a stamen that can be left on the stigma of another flower of the same species. These animal-pollinated plants produce smaller numbers of pollen grains, and their pollen grains are less likely to appear in large numbers in the environment. A few plants may use water as a transport medium for their pollen.

Regardless of how it arrives, once the pollen attaches to a female flower stigma (Figure 10.3) or a female cone of the same species, it germinates with one nucleus growing a tubelike structure that penetrates the style and grows into the ovary. This collection of cells constitutes the haploid **male gametophyte**. As the pollen tube grows toward the ovary, it brings the other haploid nucleus to the female gamete located in the **female gametophyte** or **ovule** (Figure 10.3). The second nucleus of the pollen grain divides and one of them becomes the male nucleus that will fuse with the female sex cell and develop into a diploid **embryo**. The other male nucleus fuses with one or more other female nuclei, and they develop into the endosperm. The **endosperm** will provide the energy for development until the young seedling can begin photosynthesis.

The fertilized ovum develops into an embryo that is arrested in development (see Figure 4.1). The embryo completes its early development while surrounding cells in the ovary are transformed into a seed coat containing the dormant embryo. The number of ovules and hence the number of seeds produced is determined by the genetics and resultant anatomy of each species. In angiosperms (= covered fruit), the ovary then develops into a fruit containing the seeds. In conifers, the seeds appear on the inner surfaces of the female cones (gymnosperm = naked seed).

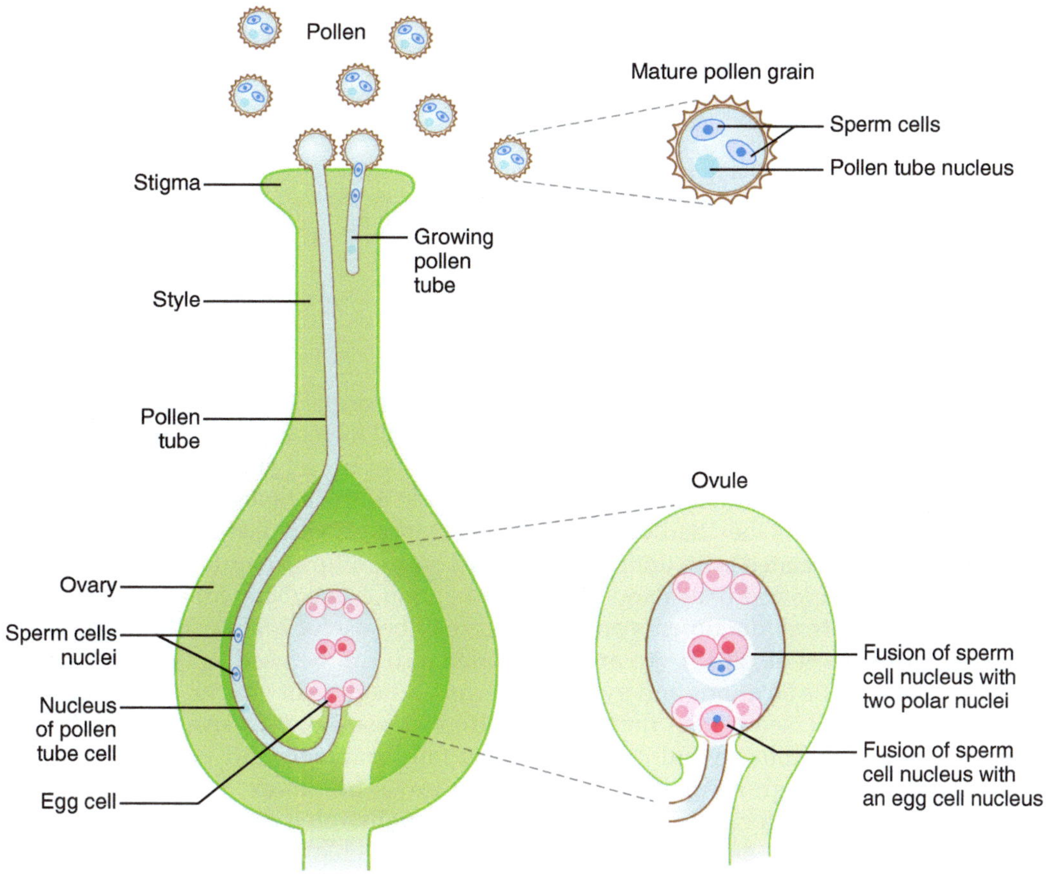

FIGURE 10.3 Pollination in a flowering plant. See text for explanation. *http://www.blog.gurukpo.com/pollination.*

1.2 Forensic Palynology

Pollen was not used forensically until the latter half of the twentieth century but since then it has found some applications in both civil and criminal cases (Table 10.1). Despite its advantages (Table 10.2), forensic use of pollen has not received an overwhelming response from the law enforcement community and the judiciary for a variety of reasons (Table 10.3). Pollen identification is labor intensive and requires considerable training and experience. It also has some difficulties with some of the Daubert criteria (Walsh and Horrocks, 2008; see Chapter 2 for Daubert criteria). The palynologist can often show that two samples are different, but when two samples are very similar in their pollen composition, can it be said they are from the same source with the exclusion of all others? Probably not, as usually there are few control sites included in forensic pollen analysis. To increase the number of comparisons would increase the time needed for the analysis and hence the cost. Walsh and Horrocks (2008) also pointed out inherent problems with establishment of error rates in pollen identification. Pollen evidence often may be only circumstantial but together with other information still may contribute substantially to an investigation and/or prosecution (Mildenhall, 2006a).

TABLE 10.3 Reasons Why Pollen Data Are Rarely Used in the US[a]

1. Inappropriate sample collection resulting in contamination of sample
2. Little legal precedence in the US courts and ignorance about palynology among investigators, lawyers, and judges
3. Lack of trained palynologists willing and able to do forensic studies
4. Concern by palynologists of liability
5. Lack of access to appropriate microscopic and/or photographic equipment
6. No major university or other training facility for pollen forensics in the US
7. Lack of available reference material for pollen identification
8. Lack of funding for forensic pollen studies

[a] *Based mostly on Bryant and Jones (2006).*

The first recorded cases using pollen in the solving of a homicide case occurred in Sweden and Austria in 1959 (see Mildenhall et al., 2006; Walsh and Horrocks, 2008). In the 1970s, pollen grains were used in the US as evidence of the source of honey (Bryant and Jones, 2006). Bees collect pollen as well as nectar and the honey made in their hives contains pollen from the species whose flowers they had visited. Pollen has since proven useful in a variety of situations and can be associated with animal fur, human hair, clothing, painted wood, rope used to tie up a victim, imported drugs and plants, to mention a few (Mildenhall, 2006a). Pollen also can be isolated from documents and ink and may be useful for identification of document fraud (Morgan et al., 2013). An array of pollen also can be used to describe the plant composition of specific locations and even to determine the time of year an event occurred.

It is the wind-dispersed pollen that lends itself most readily to forensic work. Approximately 95% of pollen from a wind-pollinated plant falls to the ground within a mile of the producing plant, although a small percentage may travel hundreds or even thousands of miles. Thus, a particular locale is characterized by the pollen that falls to the ground: often referred to as **pollen rain**. Because vegetation exhibits particular patterns on a local level as well as over large landscapes (see Figure 10.4), the analysis of pollen will reflect the vegetation present at a given site. If cores are made deeper into the soil, one can reconstruct the history of the plant communities that have occupied that site by analyzing the pollen present in different layers of the soil.

Even nearby sites may differ in the percentage of various pollen species present. Analysis of multiple surface soil samples indicates that samples within a local site are not statistically different from each other in the ratios of pollen grains of the species present but that the ratios in samples from a distant site of similar vegetation type differed significantly (Horrocks et al., 1998). Therefore, comparison of the pollen character of a surface soil sample with a soil sample from a vehicle, for example, might be a strong forensic technique linking the two.

Different species of plants in a given location produce their pollen at different times of year (Montali et al., 2006; Szibor et al., 1998). Hence, the pollen rain will differ in its composition over time. It is then possible to determine when a pollen sample was produced by its composition (Figure 10.5). Consequently, pollen associated with a victim's clothing or hair recovered from a gravesite, for example, may indicate the time of year that the person was killed (Bryant and Jones, 2006; Wiltshire, 2006). Experimental placement of corpses in the environment at different times of the year shows that they accumulate pollen from different species linked to monthly changes in the pollen rain (Figure 10.6).

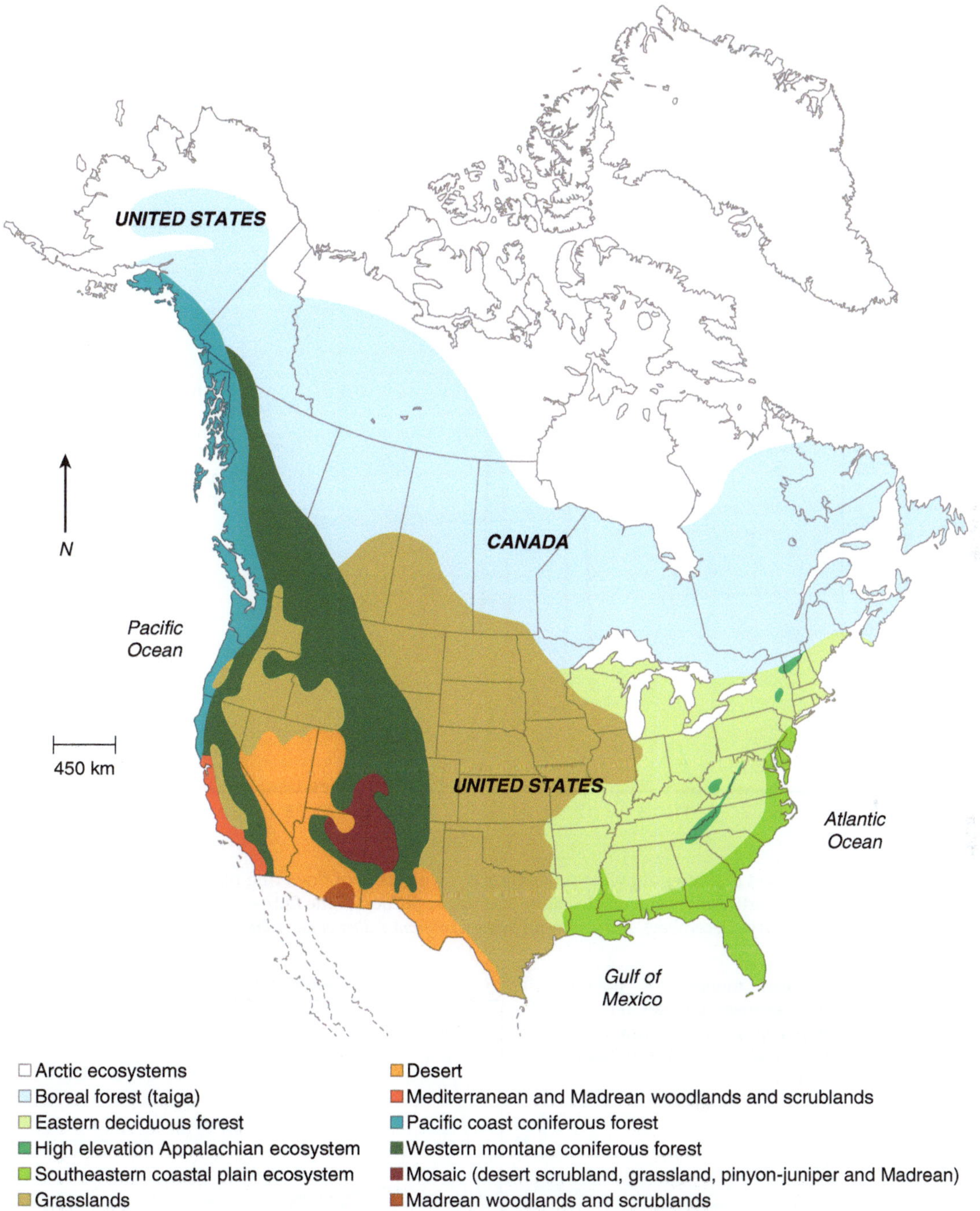

FIGURE 10.4 Vegetation zones in North America contain unique collections of plant species and hence exhibit unique pollen rain compositions. The madrean region is characterized as arid or semi-arid and having a distinctive flora. *Redrawn with permission from Bryant and Jones (2006).*

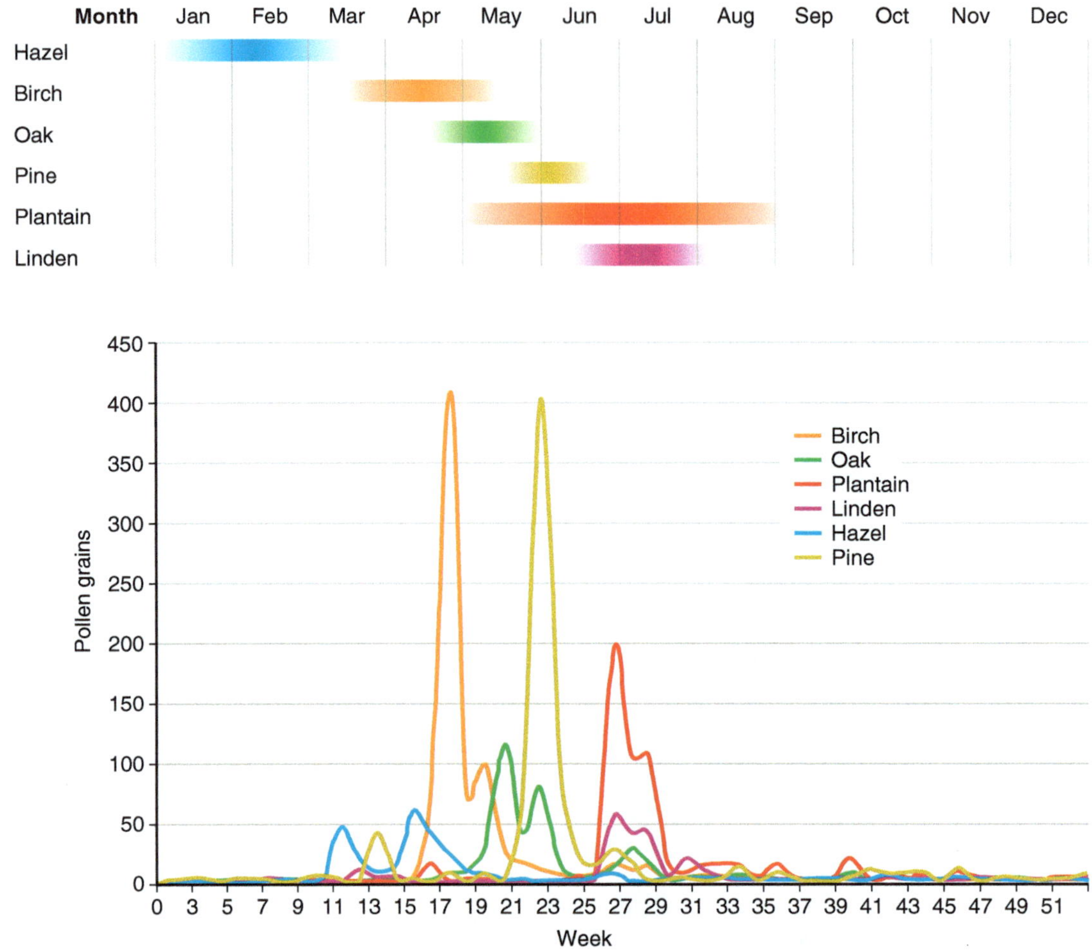

FIGURE 10.5 Variation in timing of pollen production by five genera of flowering plants and one conifer genus throughout the year in North America at one location. The upper portion of the figure depicts the intensity of pollen release by months (January–December). *Redrawn with permission from Szibor et al. (1998).*

FIGURE 10.6 Mean number of pollen grains collected from exposed corpses reflects the abundance of pollen grains captured on exposed glass slides. *Redrawn with permission from Montali et al. (2006).*

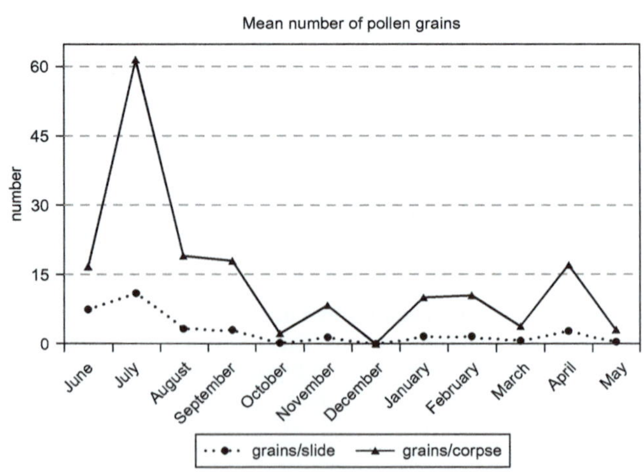

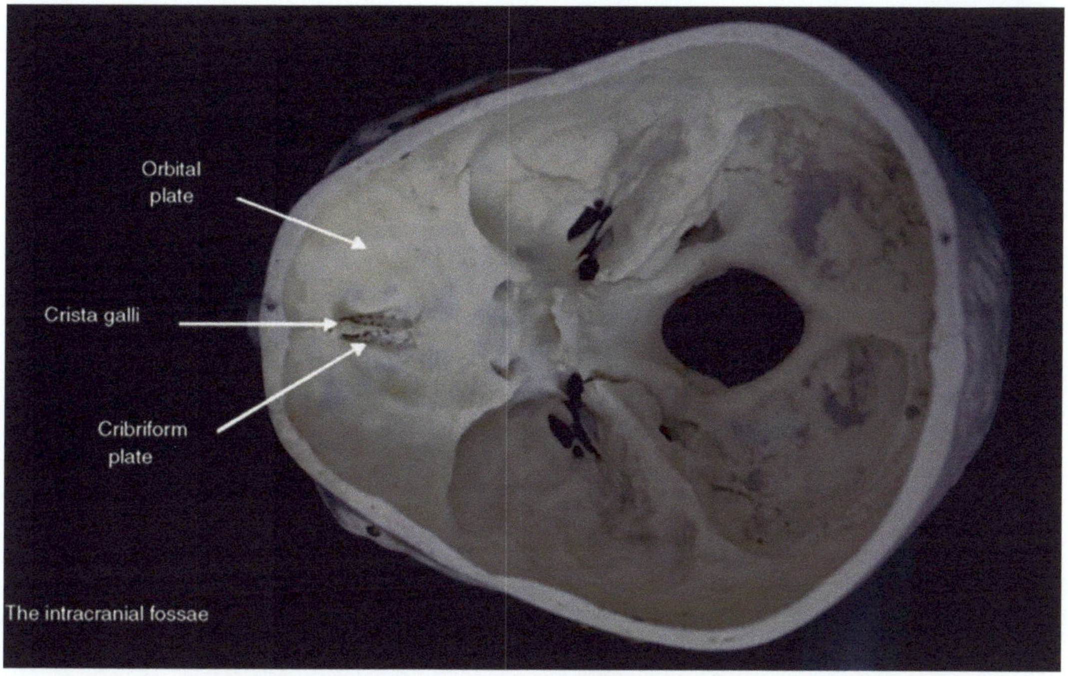

FIGURE 10.7 Location of pollen accumulation in a human skull. *Reprinted with permission from Wiltshire and Black (2006).*

Pollen also collects in people's hair, clothing, and within openings of their bodies (Bryant and Jones, 2006; Mildenhall, 2006b; Wiltshire, 2006; Wiltshire and Black, 2006). Pollen can be breathed into the sinuses and become deposited in the skull. Pollen thus trapped in the skull may indicate where and when a person died (Wiltshire and Black, 2006; Figure 10.7). Consequently, it is possible to determine at what time of year a buried person died by identifying the pollen associated with the corpse and examining the seasonal pattern of pollen release. Pollen also may be present in soil attached to a suspect's clothing or shoes (Horrocks et al., 1999) as well as associated with their vehicles (e.g., Mildenhall et al., 2006). Thus, a suspect may be connected to a crime scene through microscopic evidence of a unique pollen source that the suspect was unaware he/she was collecting.

Analysis of pollen in soil has proven useful in investigating mass graves in relation to war crimes (Brown, 2006). If bodies are exhumed and reburied at another site, the pollen pattern in the soil associated with the body may differ markedly from the second burial site soil (Brown, 2006). Thus, the pollen print in a grave may differentiate between a site where a person was killed and the site where the body was buried (Figure 10.8). Again, the pollen pattern may indicate the time of year that the crime took place (Szibor et al., 1998).

1.3 Collection and Processing of Pollen Samples

People trained in collection and uses of pollen should have early access to a crime scene since contamination with additional pollen by crime scene investigators can easily occur without their being aware of it. Failure to properly collect and process samples

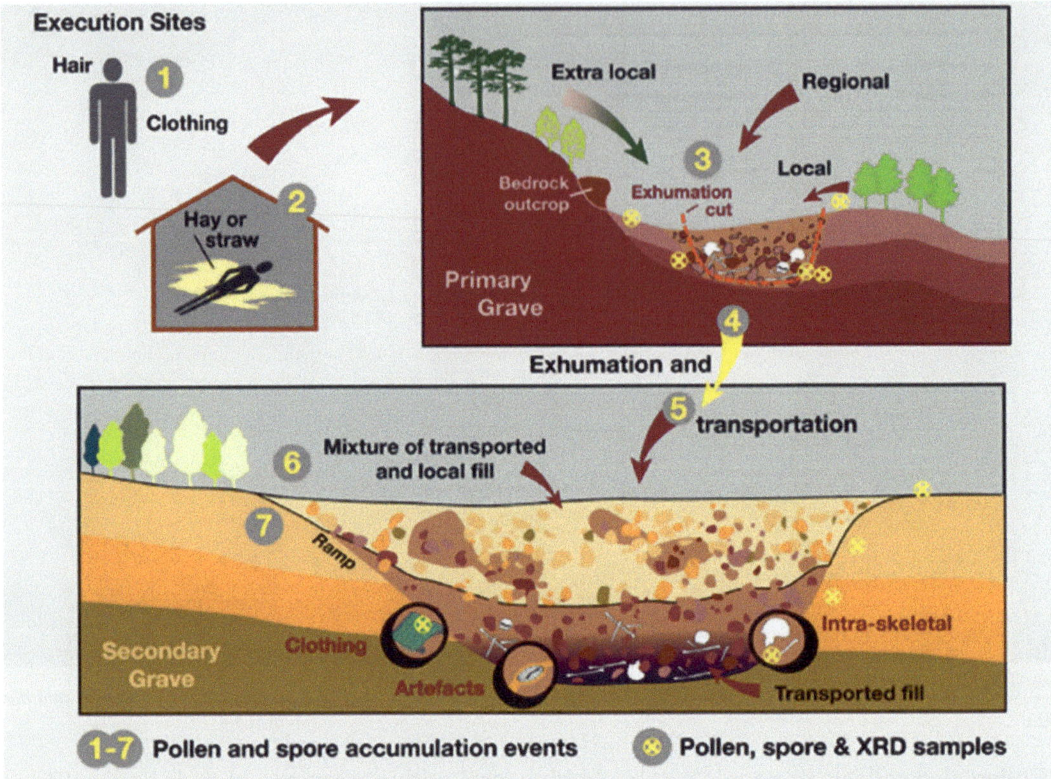

FIGURE 10.8 The pollen associated with the victims' hair and clothing identified the execution site (1,2). Following creation of a primary mass grave (3), the perpetrators decided to move the mass grave to prevent its detection (4–7). The pollen composition of the soil transported from the primary gravesite differed from the pollen pattern for the local fill used to hide the secondary gravesite. *Reprinted with permission from Brown (2006).*

may render them useless. Horrocks (2004) recommends processing the samples as soon as possible to reduce the chances of contamination by outside sources. Only when the pollen grains are mounted on slides are they safe from contamination. Preparation of the collected samples can require a considerable amount of time. Furthermore, it may take a considerable amount of time for a qualified palynologist to identify the samples, especially if scanning electron microscopy (SEM) is involved. Additionally, the association of the pollen with a given habitat may require the cooperation of plant ecologists (e.g., see Felde et al., 2014). Information on collection and processing of pollen samples can be found in several publications (especially Horrocks, 2004; see also Mildenhall et al., 2006).

1.4 Identification of Pollen

Pollen can be generally characterized with the use of high magnification with a light microscope (Figure 10.1). However, identification of a certain species may require detail only obtainable through the use of SEM that provides a detailed three-dimensional image of the pollen grains (Figure 10.2). Authenticated light microscope and SEM images for the pollen grains of numerous plant species are available at online sites (see listing at the end of the chapter). It is difficult to positively identify pollen simply from a photograph, however, and the best method for identification is to have a trained palynologist compare your samples to collections of known pollen.

1.5 Cases Involving Forensic Palynology

Pollen can be useful in determining where something came from or where a suspect has been. Even fossil pollen may be useful forensically as described here.

1.5.1 The Danube Case: Fossil Pollen Waltzes a Suspect into Jail

An Austrian man disappeared in 1959 while on a trip down the Danube River (see Mildenhall et al., 2006; Walsh and Horrocks, 2008). Police had a suspect with a motive but no evidence. However, they did obtain samples of mud from the suspect's shoes and had the mud analyzed by a palynologist. The palynologist discovered 40-million-year old hickory tree pollen in the mud that, according to geological and vegetation maps, could have come from only one site along the Danube where such ancient soil was exposed. When confronted with the precise location, the astonished suspect confessed to the crime and even showed the police where to find the body.

1.5.2 A Honey of a Case

Until the 1980s, the only records of palynology use in the US for forensic purposes involved the examination of pollen in honey during the 1970s to determine the source of the honey (Bryant and Jones, 2006). At that time, the U.S. Department of Agriculture was providing subsidies to farmers to encourage local honey production. Because bees are well known for their ability to collect pollen and return with it to their hives, honey contains pollen grains that indicate the source of the nectar collected by the bees and used to make their honey. It was possible to determine, for example, that some honey that farmers claimed to have been produced locally was actually imported from Mexico because pollen grains from species that were not present in the US were present in the Mexican honey. Thus, it was possible to determine where a food product was made and identify farmers who were buying inexpensive Mexican honey and illegally collecting the federal subsidy.

Another early use of forensic honey palynology was to settle the claim of a honey producer that a neighboring farmer using pesticides on his lima beans had damaged the bees owned by the honey producer (Bryant and Jones, 2006). The honey producer claimed his bees had contacted the spray while visiting flowers on the bean plant causing them to become sick. However, the absence of lima bean pollen in the honey produced by the bees indicated the bees had not visited the lima bean flowers and therefore were probably not directly affected by the pesticide applied by the farmer.

1.5.3 *I Didn't Go Where My Motorcycle Was Found*

In New Zealand (Mildenhall, 1990), a motorcycle involved in a robbery was chased up a wet tract of land where the rider abandoned the muddy motorcycle and fled on foot. Later, when the owner attempted to reclaim his "stolen" motorbike, he was arrested when the police noted mud clinging to his boots. The owner claimed he had gotten mud on his boots when he recently had ridden the motorcycle in a nearby tract but not where the motorcycle was found. Mud samples were collected from the suspect's boots, the motorcycle tract, the nearby "alibi" tract, and the farm where he lived. A comparison of the pollen and spore content of the mud revealed that the mud on his boots matched only one of the possible sources (see Table 10.4 and draw your own conclusion).

TABLE 10.4 Comparison of Pollen[a]/Spore Content of Mud from Four Sites Related to the New Zealand Motorcycle Case[b]

Species	Farm site	Alibi tract	Cycle tract	Suspect's boots
Pinus	Rare	Rare	Common	Common
Podocarpus	Present	Absent	Present	Present
Liverworts (spores)	Rare	Rare	Common and varied	Common and varied
Compositae family	Rare	Common	Relatively rare	Relatively rare
Caprifoliaceae family	Absent	Bottom of tract only	Present in all samples	Present in all samples
Knightia excelsa	Rare	Common	Relatively rare	Relatively rare
Gramineae	Abundant	Common	Common	Common
Cyathea (tree fern spores)	Rare	Abundant	Common	Common
Pteridium aquilinum	Rare	Rare	Abundant	Abundant

[a] *Note that in some instances it was only necessary to identify the pollen to genus or even just to family and show major differences in some of the samples from the mud on the suspect's boots.*
[b] *Based on Mildenhall (1990).*

1.5.4 *Moving a Mass Grave Takes Pollen Evidence with It*

While attempting to piece together events related to mass killings in Bosnia that occurred in 1997, it was important to identify where massacres had occurred in relation to where mass graves were uncovered (Brown, 2006). It appeared that some large primary graves had been unearthed and reburied in smaller remote sites in order to play down the extent of the major massacres. Soil present in the secondary graves close to the bodies did not match the surrounding soil characteristics. Furthermore, the pollen grains and spores associated with the bodies differed from what would be expected if the people had been killed at that site. Through soil analysis and matching of the pollen signatures it was possible to link the secondary sites to the massacre sites (Table 10.5).

TABLE 10.5 Use of Soil Type and Pollen Data to Link Primary and Secondary Gravesites in Bosnia

Primary site	Soil type/lithology/ inclusions	Major minerals	Vegetation/land use	Dominant pollen/spores (in descending value)	Linked secondary sites
Kozluk	River terrace gravels, discoid fluvial gravel (imbricated)	Ch, I/M, Qz, F, C	Scrub/grassland and arable	–	CR3
Branjevo Fm	Tuff with deep loessic soils	Ch, I/M, Ka, Qz, Fe	Edge of arable field (wheat—communal farm) ruderals	Cereals, Poaceae, *Pinus*, *Picea*	CR12
Lazete I	Thrust zone, limestone, dolomites, sandstones, serpentinite dyke	S, I/M, Ka, Qz, Fe	Edge of montane forest (10 m)	*Pinus*, Cyperaceae, Poaceae, *Picea*, *Juglans*	HZ3
Lazete II	Thrust zone, limestone, dolomites, sandstones, black water piping	S, I/M, Ka, Qz, Fe	Clearing in the montane forest, wet meadow	*Pinus*, Cyperaceae, Poaceae, *Picea*, *Juglans*	HZ2, HZ4, HZ5
Glogova 1E	Sandstones and siltstones with limestone-rich gravel in places	Qz, Ch, I/M	Mixed arable, hay meadow, orchards, and forest (including pine, beech, hornbeam, and spruce)	*Fagus*, *Picea*, *Pinus*, *Carpinus*, *Corylus*, Poaceae	ZJ6
Glogova 1F, 1H, 3, 5, 7, 8, 9	Sandstones and siltstones with limestone-rich gravel in places, hay masses (some with shell casings), rubble, concrete, plaster	Qz, Ch, I/M	Mixed arable, hay meadow, orchards, and forest (including pine, beech, hornbeam, and spruce)	Trees, *Pinus*, *Picea*, herbs (high Poaceae) and high cereals (*Avena*/*Triticum*) occasional *Zea mays*, *Malus* t., and *Prunus* t.	ZJ5

S, swelling clays; Ch, chlorite; I/M, illite/mica; F, feldspar; Ka, kaolinite; Qz, quartz; Fe, iron; C, calcite.
Reprinted with permission from Brown (2006).

2. DIATOMS

Diatoms are microscopic (mostly from 10 to 100 μm in length), unicellular, and colonial algae found in marine, brackish, and freshwaters. Their two overlapping exterior cell walls, called valves or tests, are formed of silica rather than cellulose as in higher plants, and when fitted together the resulting **frustule** protects the living tissues within. Diatoms exhibit radial or bilateral symmetry (Figures 10.9 and 10.10). It is estimated that there are as many as 100,000 species of diatoms, although only about 8000 species have been described. Diatoms are very important photosynthetic organisms in aquatic and marine systems. Different bodies

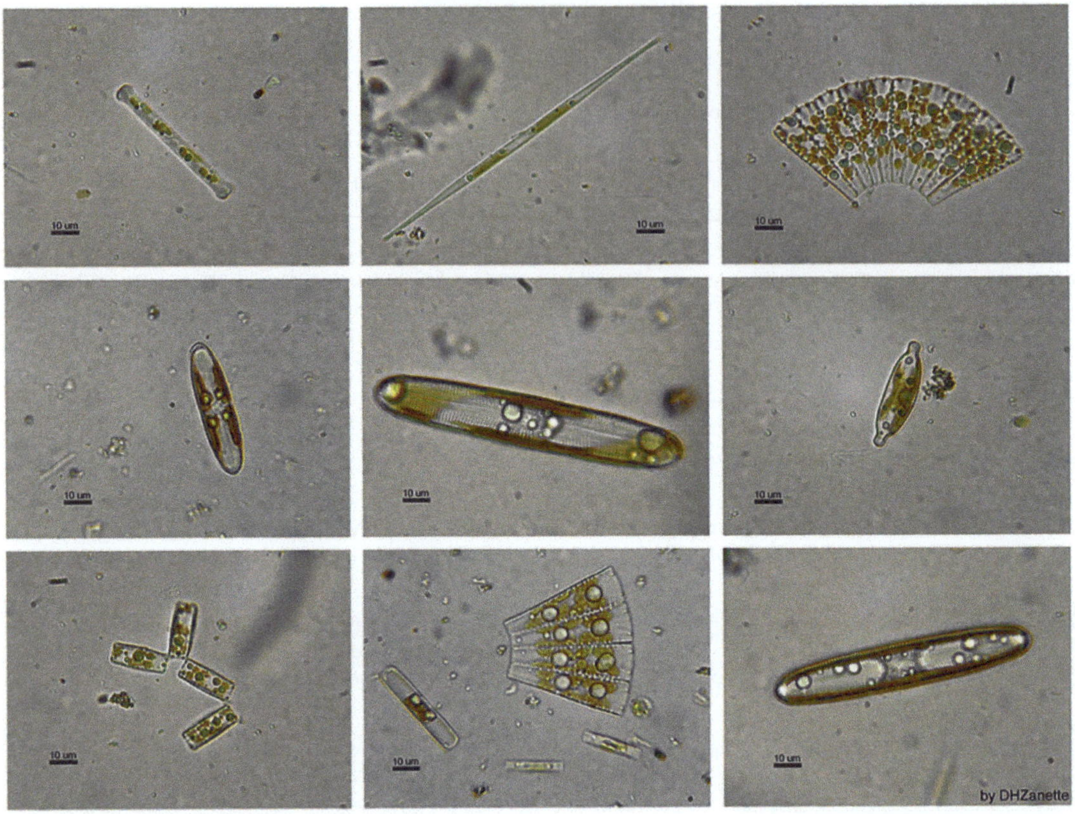

FIGURE 10.9 Examples of freshwater diatoms including two colonial forms as seen with a light microscope. *Diatoms: eukaryotic algae, courtesy of Damian H. Zanette, available at: http://commons.wikimedia.org/wiki/File:Diatomeas_w.jpg.*

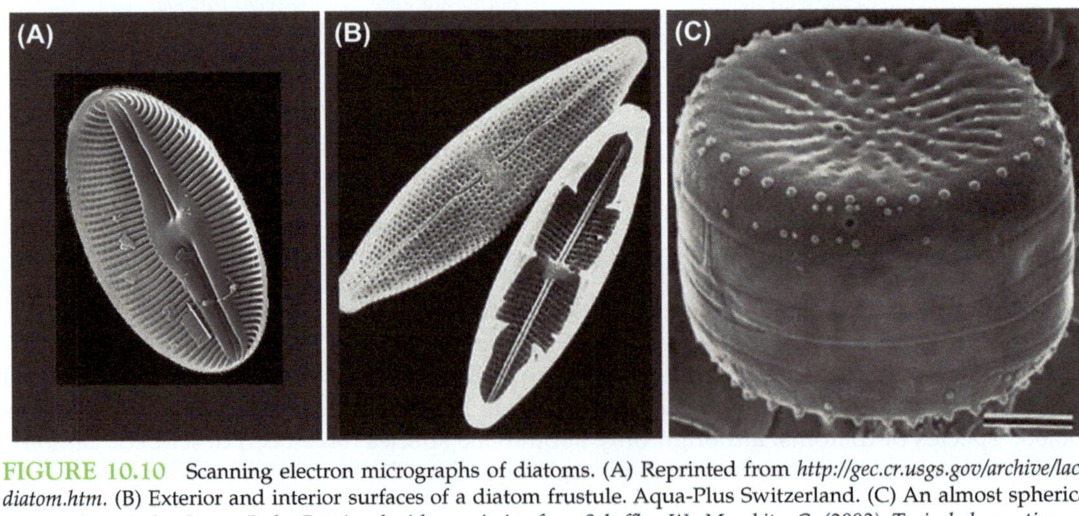

FIGURE 10.10 Scanning electron micrographs of diatoms. (A) Reprinted from *http://gec.cr.usgs.gov/archive/lacs/diatom.htm.* (B) Exterior and interior surfaces of a diatom frustule. Aqua-Plus Switzerland. (C) An almost spherical diatom from Lake Como, Italy. *Reprinted with permission from Scheffler, W., Morabito, G. (2003). Topical observations on centric diatoms (Baccillariophyceae, Centales) of Lake Como, Italy. Journal of Limnology 62, 47–60.*

of water may contain very different combinations of species and/or unique species, and they can sometimes be used to identify the aquatic source from which they came. Because of their silica frustules, diatoms are often added to cleaners, polishes, and other products as an abrasive. Ancient deposits of marine diatoms, called diatomaceous earth, are often mined for that purpose.

When a person drowns in freshwater, diatoms are taken into the lungs with the water and are distributed via blood vessels to other internal organs (Verma, 2013; Figure 10.11). Their presence in lungs, stomach, or even in bone marrow is used as evidence that the person was alive when they entered the water (**antemortem immersion**). The absence of internalized diatoms or the failure to match the internalized diatoms with those of the recovery site would indicate that the victim possibly was dead due to another cause prior to insertion into the water (**postmortem immersion**) or was drowned elsewhere, respectively. It is also possible that a person did not internalize any diatoms during drowning.

Diatoms may be extracted from the femur of a corpse and used to verify death by drowning, especially when many of the soft tissues have disappeared from the corpse (Pollanen, 1998; Pollanen et al., 1997; Vinayak et al., 2013). If the victim also swallows water, diatoms may be found in the stomach contents as well. However, the presence of diatoms (for example, from cleaning products) in lungs and other tissues of people dying from other causes has been documented (see Kumar et al., 2012). Hence, corroborative environmental data (for example, ratios of the same species in a relevant body of water) are important factors to be considered.

When a body is immersed in water (ante- or postmortem), diatoms become associated with clothing, skin, and hair (Vinayak et al., 2013). The presence of certain species can provide evidence of the time that the body was in the water prior to its discovery (**postmortem submersion interval** (PMSI)). Experimental immersion of stillborn pig carcasses in a brackish pond demonstrated a decrease over time in the number of diatom species attached to the body (Zimmerman and Wallace, 2008). Conversely, Casamatta and Verb (2000) observed an increase in the number of diatom species following immersion of rat carcasses in a river environment. The rapid degradation of the pig carcasses in 2 weeks following immersion in the brackish water pond perhaps led to diminished sites for diatom attachment. The rat carcasses exhibit about 20 different species after 1–2 weeks of immersion and as many as 50 species by day 31. Obviously, much more research is needed before these differences in colonization by diatoms can be reconciled and the data applied to the human condition for calculating a PMSI. If a human victim was determined to have drowned at that site where the body was found, information on PMSI also may be helpful in establishment of the postmortem interval.

2.1 Collection and Preparation of Diatoms from Bodies of Water

Tutorials for collection of diatoms from the relevant environment as well as procedures for preservation and mounting of diatoms on slides are readily available online (e.g., Franchini, 2013; Sterrenberg: http://micrap.selfip.com:81/micrapp/cleandiatoms.pdf). A video tutorial is also available (https://www.youtube.com/watch?v=Cp9ym5M0RUc). Note especially Sterrenberg's procedure for mounting the diatoms on the coverslip rather than directly on the slide. This is the standard procedure for mounting diatoms. Methods for extracting diatoms from clothing have been described as well (Uitdehaag et al., 2010; Scott et al., 2014).

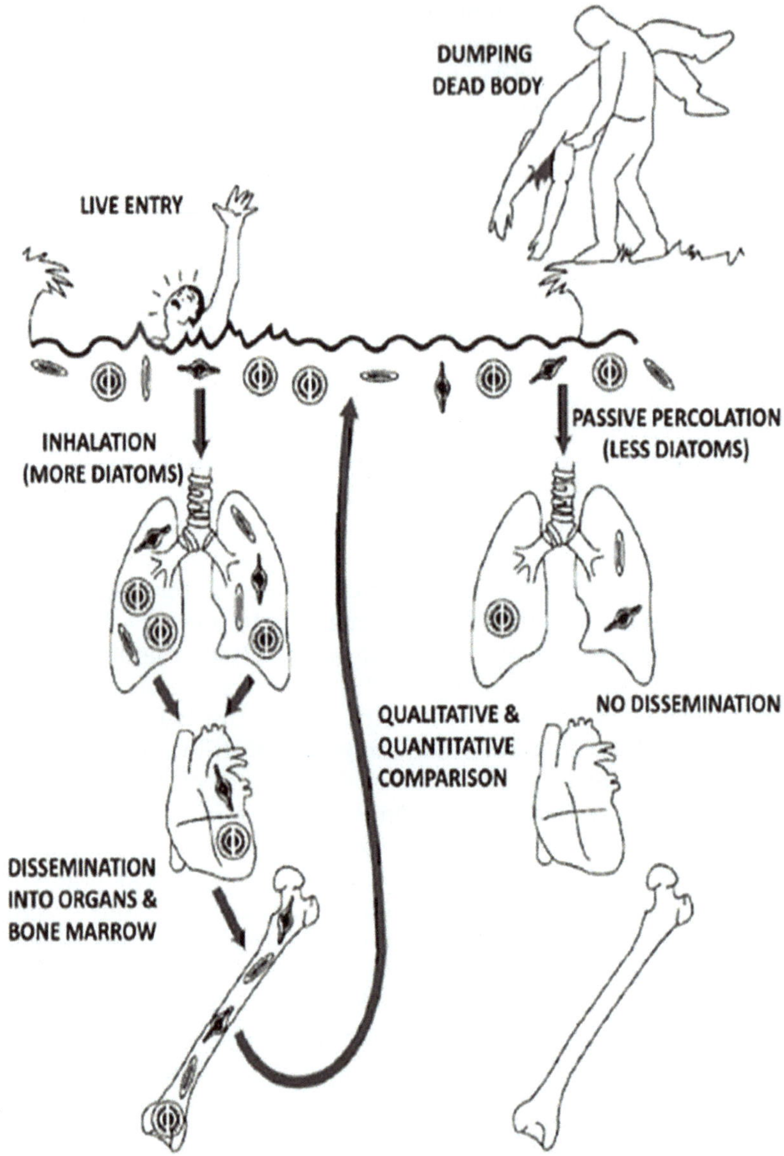

FIGURE 10.11 Differences in diatom distribution among internal organs between a live drowning (antemortem immersion) and dumping of a dead person (postmortem immersion) into a body of water. *Modified with permission from Anu and Resmi (2014).*

2.2 Cases Involving Forensic Diatomology

Numerous cases of possible drowning have used the presence and identification of diatoms to confirm whether death occurred by drowning or before the victim entered the water. Diatoms also have been used to estimate how long a body has been in the water. In other cases, diatoms may link a victim or a suspect to a particular location.

2.2.1 *Determination of Deaths due to Drowning*

Often it is critical to determine if a victim was dead before they were placed in a body of water, if they were thrown into the water or were held under the surface and they drowned, or if they accidentally drowned. Numerous cases have been reported from India, for example, where the presence of diatoms was used in determining some of these questions. Malik et al. (2013) have summarized several cases. In some cases, drowning was confirmed as the cause of death by the presence of diatoms in bones. In others, it was determined that the victim had died before the body was immersed in the water.

2.2.2 *Diatoms in an Infant's Stomach Draw Out a Confession*

A teenage mother in northern California claimed her infant son was abducted by a group of boys while she and the baby boy were playing near a fountain in a city park. The infant's body was discovered floating face down in a nearby stream, an apparent victim of drowning. The mother claimed she had not been at the stream with the baby although some witnesses said they had seen her at both the fountain and the stream. The father of the child was no longer involved with the mother and child.

Being suspicious of her story, investigators sent us samples of water from the fountain and from the river as well as samples from his stomach. Diatoms were found in the samples from the fountain and the stream as well as in the child's stomach. However, diatoms from both sources were present in the child's stomach sample (Figure 10.10). We sent photographs of the various diatoms we found to the California police department and had them set up a demonstration to see if naïve observers could match up the samples as we had. They obtained 100% agreement. When presented with the evidence that the baby had ingested water from both the fountain and the stream, the teenage mother admitted attempting to drown the child in the fountain and then, believing he was already dead, threw him into the stream, though apparently still alive. Then she concocted the story of abduction. At the time of the murder, she was pregnant with the child of another man.

2.2.3 *A New England Misadventure*

In New England during the summer of 1991, two boys who were fishing in a suburban pond were attacked by several teenagers who bound them with duct tape, beat them with a baseball bat, and dragged them into the pond and left them to drown. One boy got free, rescued his comrade, and summoned help. Later, three suspects were identified. Their sediment-encrusted sneakers were shown to exhibit the same species of diatoms and chrysophytes (another category of planktonic algae) as found at the pond, and they were found in the same proportions. This was sufficient to implicate them in the crime (see Siver et al., 1994).

References

Anu, S., Resmi, S., 2014. Forensic diatomology. Health Sciences 2014, 1–10.

Bhattacharya, K., 2011. A Textbook of Palynology, third ed. New Central Book Agency.

Brown, A.G., 2006. The use of forensic botany and geology in war crimes investigations in NE Bosnia. Forensic Science International 163, 204–210.

Bryant, V.M., 2015. Pollen. In: Presented at the Texas State Historical Society Annual Meeting in 2015. http://www.tshaonline.org/handbook/online/articles/sop.

Bryant, V.M., Jones, G.D., 2006. Forensic palynology: current status of a rarely used technique in the United States of America. Forensic Science International 163, 183–197.

Casamatta, D.A., Verb, R.G., 2000. Algal colonization of submerged carcasses in a mid-order woodland stream. Journal of Forensic Sciences 45, 1280–1285.

Felde, V.A., Bjune, A.E., Grytnes, J.-A., Birk, H.J.B., 2014. A comparison of novel and traditional numerical methods for the analysis of modern pollen assemblages from major vegetation landform types. Review of Palaeobotany and Palynology 210, 22–36.

Franchini, W., 2013. The Collecting, Cleaning and Mounting of Diatoms. Modern Microscopy How to Tutorial Series. http://www.modernmicroscopy.com/main.asp?article=107.

Horrocks, M., 2004. Sub-sampling and preparing forensic samples for pollen analysis. Journal of Forensic Sciences 49, 1–4.

Horrocks, M., Coulson, S.A., Walsh, K.A.J., 1998. Forensic palynology: variation in the pollen content of soil surface samples. Journal of Forensic Sciences 43, 320–323.

Horrocks, M., Coulson, S.A., Walsh, K.A.J., 1999. Forensic palynology: variations in the pollen content of soil on shoes and in shoeprints in soil. Journal of Forensic Sciences 44, 119–122.

Kumar, M., Deshkar, J., Naik, S.K., Yadav, P.K., 2012. Diatom test- past, present and future: a brief review. The Indian Journal of Research and Reports in Medical Sciences 2, 28–32.

Malik, M.K., Pooja Jakhar, P., Kadian, A., 2013. Role of diatoms in forensic investigation: case studies from Haryana. International Journal of Forensic Science & Pathology. 1 (3), 11–12. http://scidoc.org/IJFP-01-302.php.

Mildenhall, D.C., 1990. Forensic palynology in New Zealand. Review of Paeleobotany and Palynology 64, 227–234.

Mildenhall, D.C., 2006a. *Hypericum* pollen determines the presence of burglars at the scene of a crime: an example of forensic palynology. Forensic Science International 163, 231–235.

Mildenhall, D.C., 2006b. An unusual appearance of a common pollen type indicates the scene of a crime. Forensic Science International 163, 236–240.

Mildenhall, D.C., Wiltshire, P.E.J., Bryant, V.M., 2006. Forensic palynology: why do it and how it works. Forensic Science International 163, 163–172.

Milne, L.A., Bryant Jr., V.M., Mildenhall, D.C., 2005. Forensic palynology. In: Coyle, H.M. (Ed.), Forensic Botany: Principles and Applications to Criminal Casework. CRC Press, pp. 217–252.

Morgan, R.M., Davies, G., Balestri, F., Bull, P.A., 2013. The recovery of pollen evidence from documents and its forensic implications. Science and Justice 53, 375–384.

Montali, E., Mercuri, A.M., Grande, G.T., Accorsi, C.A., 2006. Towards a "crime pollen calendar"—Pollen analysis on corpses throughout one year. Forensic Science International 163, 211–223.

Pollanen, M.S., Cheung, C., Chiasson, D.A., 1997. The diagnostic value of the diatom test for drowning 1. Utility: a retrospective analysis of 771 cases of drowning in Ontario, Canada. Journal of Forensic Science 42, 281–285.

Pollanen, M.S., 1998. Forensic Diatomology and Drowning. Elsevier, Amsterdam, The Netherlands. pp. 1–157.

Scott, K.R., Morgan, R.M., Jones, V.J., Cameron, N.G., 2014. The transferability of diatoms to clothing and the methods appropriate for their collection and analysis in forensic geoscience. Forensic Science International 241, 127–137.

Siver, P.A., Lord, W.D., McCarthy, D.J., May 1994. Forensic limnology: the use of freshwater algal community ecology to link suspects to an aquatic crime scene in southern New England. Journal of Forensic Sciences, JFSCA 39, 847–853.

Szibor, R., Schubert, C., Schöning, R., Krause, D., Wendt, U., 1998. Pollen analysis reveals murder season. Nature 395, 449–450.

Uitdehaag, S., Dragutinovic, A., Kuiper, I., 2010. Extraction of diatoms from (cotton) clothing for forensic comparisons. Forensic Science International 200, 112–116.

Verma, K., 2013. Role of diatoms in the world of forensic science. Journal of Forensic Research. 4, 1–4. http://dx.doi.org/10.4172/2157-7145.1000181.

Vinayak, V., Mishra, V., Goyal, M.K., 2013. Diatom fingerprinting to ascertain death in drowning cases. Journal of Forensic Research 4, 207–212. http://dx.doi.org/10.4172/2157-7145.1000207.

Walsh, K.A.J., Horrocks, M., 2008. Palynology: its position in the field of forensic science. Journal of Forensic Science 53, 1053–1060.

Wiltshire, P.E.J., 2006. Hair as a source of forensic evidence in murder investigations. Forensic Science International 163, 241–248.

Wiltshire, P.E.J., Black, S., 2006. The cribriform approach to the retrieval of palynological evidence from the turbinates of murder victims. Forensic Science International 163, 224–230.

Zimmerman, K.A., Wallace, J.R., 2008. The potential to determine a postmortem submersion interval based on algal/diatom diversity on decomposing mammalian carcasses in brackish ponds in Delaware. Journal of Forensic Sciences 53, 935–941.

Further Reading

Dommelen, J.V., 2005. Limnology. Forensic Botany. Dalhousie University.
Horton, B.P., 2007. Diatoms and forensic science. Paleontological Society Papers 14–22.
Sterrenburg, F.A.S. How to Prepare Diatom Samples. http://micrap.selfip.com:81/micrapp/cleandiatoms.pdf (accessed 28.08.15).

Pollen image sources

U.S. Department of Agriculture. http://pollen.usda.gov/Photographs.htm.
University of Arizona. http://www.geo.arizona.edu/palynology/polonweb.html.
University of Wisconsin. http://www.geography.wisc.edu/faculty/williams/lab/Downloads.html.
State of California. http://oag.ca.gov/cci/reference/pollen.

CHAPTER

11

Summation and a Look to the Future

Forensic botany is an underutilized resource for forensic science that needs to be brought to the consciousness of law enforcement personnel and forensic scientists. It is simple, inexpensive for the most part, and readily explainable in the courtroom. The methods employed are supported by hundreds of years of careful scientific observations and the methods and results are readily explainable to the jurors and even the average lawyer. Although the perspective of this book is based primarily on our forensic work done for the prosecution of criminal cases, these approaches are directly applicable to the defense as well.

One of the side benefits we have received from our forensic work has been the discovery of so many dedicated and competent people working in various aspects of law enforcement. The media makes a lot of noise whenever there is a hint of misconduct but they rarely chronicle the thousands of people who are dedicated to true justice. The elements of the American justice system we have come to know include judges, defense and prosecuting attorneys and their aides, law enforcement officials (especially police and sheriff officers, state and federal agents), juries, and other forensic scientists. Within each of these categories, we have encountered some less than honorable people. And the bad ones tend to get lots of publicity. There are lawyers who seek high profile cases to improve their chances of obtaining greater political power and forensic people who falsify laboratory results believing they are destined to rid the streets of criminals. There are some forensic scientists who believe they are infallible and that their judgments are not to be questioned. Additionally, there are others willing to sell their testimonies to the highest bidder. But the overwhelming majority of the people working in the criminal justice system today are hard working, honest, and dedicated to justice.

1. A BRIEF SUMMARY OF FORENSIC PLANT SCIENCE APPROACHES

We have discussed the history of forensic plant science (Chapter 1) and have attempted to provide an overview of forensic approaches employing botanical materials with a focus on **plant anatomy, taxonomy**, and **ecology** (Chapters 4, 6, 8, and 10). In addition, we have provided examples of actual cases using these botanical approaches, mostly cases in which we were involved (Chapters 5, 7, 9, and 10).

1.1 Plant Anatomy

Plant anatomy focuses on the microscopic examination of cells and tissues. This is especially useful for the examination of **gastrointestinal (GI) contents** as well as former GI contents such as **vomitus** and **feces** (see Chapters 4 and 5). In many instances, specific plants can be identified whereas in other situations only similar cells or tissues can be documented. It is important when analyzing stomach samples to know the composition of the victim's last known meal and to see if the material in the stomach is consistent with the anatomy of things present in that last meal even if it is only possible to say, yes, this is a cell from the edible pulp of a bean and not identify which bean (e.g., kidney bean vs navy bean). Similarly, plant materials are more dissociated in fecal samples than in stomach contents, and there are many fragments that cannot be linked directly to a specific species. Nevertheless, they can be used to match two samples with respect to content and frequency of appearance. Such examinations can be helpful in determining time of death (e.g., stomach contents) as well as for connecting suspects to a specific primary or secondary crime scenes (e.g., feces).

1.2 Plant Taxonomy

Taxonomy involves the identification of species primarily using morphological characteristics that have strong genetic bases. Often a species can be identified from only a fragment of the parent plant. For example, a particular maple tree may be identified from only a leaf or a fruit or a twig. Local **dichotomous keys** are generally available to help identify plants.

Identification of a species of plant is very important in drug cases or cases of plant-related poisonings. Identification of a unique plant such as a rare or exotic species may link a suspect, a victim, or their possessions, such as a vehicle or clothing, to a specific location. Taxonomy also is a central element in forensic plant ecology (see ahead).

The microscopic examination of **pollen** or **spores** to identify species (see Chapter 10) also can be useful but has some important weaknesses when compared to morphological identification of species. The first weakness is that it is not always possible to identify a particular species and the analysis may only identify the genus or family. The second drawback is the difficulty in finding a qualified expert to undertake the analysis. Furthermore, the process may take much longer and be more expensive than working with a morphological taxonomist.

Phytoliths, the silicon dioxide ghosts of the living plant cells, also can be used to identify species of plants (Piperno, 2006). Like pollen, phytoliths can persist in soil for thousands of years after the plants have died. These structures have been useful in determining what types of plants were growing there in the past and provide clues as to what type of climate existed at that location. They also can be used to compare soil samples associated with a crime. However, like the situation for pollen analyses, there are few scientists available with expertise to identify phytoliths. Furthermore, identification may be diagnostic only of genus or family in many cases. They are especially useful in the identification of fossil grasses (Poaceae), however.

1.2.1 Use of DNA for Species Identification

One might ask why not simply use DNA techniques to identify plant species? Utilization of DNA analyses for identification of plant species and especially of individual plants is in its infancy as compared to DNA analyses of humans (see Chapter 3 for details). Thus, comparisons are still experimental in nature and more costly to perform.

Techniques for the analyses of **nuclear**, **mitochondrial (mtDNA)**, or **chloroplast DNA (cpDNA)** for identification of plant species are being developed but for only a few of the hundreds of thousands of plant species, many of which have yet to be described. **The Plant List** (http://www.theplantlist.org) contains over 1 million described species of which about 33% are established as valid. There are possibly about 200,000 species of **diatoms** as well as approximately 73,000 other described species of **algae** (Guiry, 2012). The number of described species of **fungi** is about 100,000 (http://www.infoplease.com/encyclopedia/science/fungi-types-fungi.html). The Plant List includes about 20,000 accepted species of **bryophytes** (nonvascular land plants including mosses, liverworts, and hornworts) although many others have been proposed. There are about 13,000 described species of **ferns**, representing the most primitive group of vascular plants. **Gymnosperms** consist of only about 1000 species, many of which are **conifers**. Although relatively small in numbers of species, conifers are very important plants because they dominate many northern latitude ecosystems. The most advanced group of plants, the **angiosperms** or flowering plants, consists of more than 300,000 verified species that include most of the plants we consume as food. Identification of virtually all of these angiosperm species is based entirely on morphological criteria and the genomes have been sequenced for only a handful of angiosperms (e.g., less than 0.003% of the flowering plants) and a few representatives from other plant groups (see Table 3.1).

It appears that the best method available today for genetic identification of plant species is **genetic barcoding** using cpDNA. Initially, cpDNA appeared to have too little variation to be useful taxonomically (Chase et al., 2005). However, recent studies are developing an international standard for cpDNA taxonomy of terrestrial plants (see reviews by Ajmal Ali et al., 2014; Ferri et al., 2015). This approach has proven useful already for identification of poisonous plants (Bruni et al., 2010; see Table 11.1), although barcoding will not distinguish among individuals of a given species.

The technique of **amplified fragment length polymorphism** markers may prove useful for identifying plant species whose nuclear genomes are not known (see Chapter 3). Meanwhile, the technique of barcoding employing cpDNA seems to be the most promising approach for future taxonomic use although barcoding has not always separated some plant species successfully. As yet, few laboratories are doing this type of analysis, and it is unlikely that commercial kits will appear any time soon to facilitate identification in the forensic laboratory. Until then, identification by classical taxonomic techniques will be the most expedient and least expensive approach to species identification.

1.2.1.1 DNA AND THE IDENTIFICATION OF INDIVIDUAL PLANTS

A few studies have typed individual plants successfully using nuclear DNA (see Chapter 3), and there certainly is the possibility of identifying individuals in the future as genetic techniques improve. However, Mother Nature and the genetic breeding of agricultural varieties in the laboratory have provided some additional complications.

Selective breeding of plants by humans over millennia makes it more difficult to identify individual plants of a given species or variety. When two plants exhibit identical forms of the same gene (i.e., the same allele), we say they are **homozygous** for that gene. Selective inbreeding for certain desirable characteristics often results in considerable homozygosity of

TABLE 11.1 Poisonous Plants Identified through Chloroplast DNA Analyses and Toxic Features[a]

Species	Family	Common name	Poisonous organ	Toxic substance(s)
Nandina domestica Thunb.	Berberidaceae	Sacred bamboo	Fruit and other part	Hydrocyanic acid and nandenine
Ilex aquifolium L.	Aquifoliaceae	Holly	Fruit, leaves, and seed	Theobromine, alkaloid, and glucoside
Aucuba japonica Thunb.	Garryaceae	Spotted laurel	Fruit, leaves	Aucubin and different glycosides
Arum italicum Mill.	Araceae	Lords-and-ladies	All parts	Calcium oxalate crystals
Arum maculatum L.	Araceae	Lords-and-ladies	All parts	Calcium oxalate crystals
Convallaria majalis L.	Ruscaceae	Lily of the valley	All parts	Cardiac glycosides and saponins
Ruscus aculeatus L.	Ruscaceae	Butcher's broom	Fruit	Unknown
Hedera helix L.	Araliaceae	Common ivy	All parts	Triterpenoid saponins and polyacetylene
Hedera hibernica (G.Kirchn.) Bean.	Araliaceae	Irish ivy	All parts	Triterpenoid saponins and polyacetylene
Ligustrum vulgare L.	Oleaceae	European privet	Berries	Ligustrin, syringin, and other glycosides
Ligustrum lucidum W.T. Aiton	Oleaceae	Glossy privet	Berries	Ligustrin, syringin, and other glycosides
Ligustrum japonicum Thunb.	Oleaceae	Japanese privet	Berries	Ligustrin, syringin, and other glycosides
Phytolacca americana L.	Phytolaccaceae	American pokeweed	All parts	Phytolaccatoxin and related triterpene saponins, alkaloid, and histamines
Ficus benjamina L.	Moraceae	Weeping fig	Plant sap from all parts	Furocoumarins, psoralens, ficin
Monstera deliciosa Liebm.	Araceae	Mexican breadfruit	All parts	Needlelike calcium oxalate crystals and other unidentified toxins
Philodendron sp.	Araceae	Philodendron	All parts	Calcium oxalate crystals and other toxins
Dieffenbachia seguine (Jacq.) Schott	Araceae	Dumb cane	All parts	Calcium oxalate crystals, oxalic acid
Spathiphyllum wallisii Regel	Araceae	Peace lily	Leaves	Calcium oxalate crystals
Trachelospermum jasminoides Lem.	Apocynaceae	Star jasmine	Leaves	Unknown
Schefflera arboricola (Hayata) Merr.	Araliaceae	Schefflera	Leaves, plant sap	Oxalates

TABLE 11.1 Poisonous Plants Identified through Chloroplast DNA Analyses and Toxic Features—cont'd

Species	Family	Common name	Poisonous organ	Toxic substance(s)
Sansevieria trifasciata Prain	Ruscaceae	Snake plant	All parts	Possibly saponins and organic acids
Hydrangea macrophylla (Thunb.) Ser.	Hydrangeaceae	Hydrangea	All parts	Cyanogenic glycoside such as hydrangin
Wisteria sinensis (Sims) Sweet	Fabaceae	Chinese wisteria	Seed and other parts	Glycoside (i.e., wisterin) and a toxic resin
Nerium oleander L.	Apocynaceae	Oleander	All parts, green or dry	Glycosides, nerioside, oleandroside, etc.
Skimmia reevesiana (Fortune) Fortune	Rutaceae	Skimmia	Fruit	Alkaloid called "skimmianin"
Kalanchoë daigremontiana Hamet & Perrier	Crassulaceae	Mexican hat plant	Leaves, stems	Glycoside such as daigremontianin
Anthurium andraeanum Linden	Araceae	Tail flower	All parts	Calcium oxalate crystals
Veratrum lobelianum Bernh.	Melanthiaceae	False helleborine	All parts	Steroidal alkaloids
Veratrum nigrum L.	Melanthiaceae	Black false helleborine	All parts	Steroidal alkaloids
Lycianthes rantonnetii (Carrière) Bitter	Solanaceae	Blue potato bush	All parts	Different alkaloids
Atropa belladonna L.	Solanaceae	Deadly nightshade	All parts, mainly berries	Tropane alkaloids, atropine, and others
Colchicum autumnale L.	Colchicaceae	Meadow saffron	All parts	Alkaloid colchicine
Aconitum lycoctonum L.	Ranunculaceae	Wolfs bane	All parts	Alkaloids, aconitine, and others
Aconitum napellus L.	Ranunculaceae	Monkshood	All parts	Alkaloids, aconitine, and others
Aconitum degenii Gáyer subsp. *paniculatum* (Arcang.) Mucher	Ranunculaceae	Panicled Monkshood	All parts	Alkaloids, aconitine, and others
Aconitum anthora L.	Ranunculaceae	Pyrenean Monkshood	All parts	Alkaloids, aconitine, and others
Sambucus ebulus L.	Adoxaceae	Dwarf elder	Fruit and other parts	Cyanogenic glycoside and others
Sambucus racemosa L.	Adoxaceae	Red-berried elder	Edible fruit	Cyanogenic glycoside in vegetative parts
Sambucus nigra L.	Adoxaceae	Elder	Edible fruit	Cyanogenic glycoside in vegetative parts
Solanum dulcamara L.	Solanaceae	Bittersweet	Berries	Solanine and other alkaloids

(Continued)

TABLE 11.1 Poisonous Plants Identified through Chloroplast DNA Analyses and Toxic Features—cont'd

Species	Family	Common name	Poisonous organ	Toxic substance(s)
Solanum nigrum L.	Solanaceae	Black nightshade	Berries	Solanine and other alkaloids
Solanum pseudocapsicum L.	Solanaceae	Jerusalem cherry	Fruits and other parts	Alkaloids such as solanocapsine
Prunus armeniaca L.	Rosaceae	Apricot	Edible fruit	Seeds contain cyanogenic glycoside
Prunus avium L.	Rosaceae	Cherry	Edible fruit	Seeds contain cyanogenic glycoside
Prunus cerasus L.	Rosaceae	Sour cherry	Edible fruit	Seeds contain cyanogenic glycoside
Prunus domestica L.	Rosaceae	Plum	Edible fruit	Seeds contain cyanogenic glycoside
Prunus laurocerasus L.	Rosaceae	Cherry laurel	Vegetative parts, fruit, and seed	Cyanogenic glycoside, amygdalin, and others
Prunus persica (L.) Batsch	Rosaceae	Peach	Edible fruit	Seeds contain cyanogenic glycoside

[a]*Modified with permission from Bruni et al., (2010).*

the entire genome in particular varieties of crop species, both ornamental plants and those used for food. Furthermore, seed companies produce uniform hybrid varieties such that all of the plants that develop from those seeds are genetically alike and hence produce a uniform crop. Thus, separate fields with the same form of hybrid corn may be genetically identical even though they are grown hundreds of miles apart.

The absence of genetic variation can occur in wild plants as well although on a smaller scale. One of the peculiarities of plant reproduction that is uncommon for animals is the occurrence of asexual or **vegetative reproduction** that can result in production of clones that are genetically identical. For example, many plants may send out branches that contact the soil and sprout new roots. This new growth may later separate from the parent, becoming an independent clone. Other species may sprout new growths from lateral roots, and these new shoots appear to be separate plants although they still may be connected underground to the parent plant. Others may produce asexual **propagules** that bud off from the parent plant and develop into clones. Thus, an area that appears to be inhabited by a number of individual plants may in fact all be clones derived from a single parent and have an identical genome as that parent. Examination of all the trees in a single stand of aspens in the mountains revealed that it consisted of 47,000 connected individual clones covering 43 ha, making it appear to be the world's largest organism (see Grant et al., 1992).

1.2.1.2 IDENTIFICATION OF GENETICALLY MODIFIED ORGANISMS

Genetically modified organisms (GMOs) are produced by inserting genetic material (sometimes from another species) into a plant such that the new genetic material will provide the plant the ability to exhibit some desirable trait (i.e., genetic engineering). It essentially is a short cut to the traditional genetic breeding and selection experiments that have resulted in the varieties of food animals and plant crops we now produce for human consumption. Sometimes this genetic modification leads to undesired consequences. Many scientists and nonscientists are opposed to this technology for a variety of reasons. Development of genetically engineered pesticide-resistant crops has resulted in the passage of these "pesticide genes" to pest species and resulted in development of pesticide-resistant **superweeds** as well as increased contamination of the environment with the pesticide chemicals that have deleterious effects on useful insects such as pollinator species.

The development of GMOs in agriculture has spawned numerous movements to limit development of GMOs or to have products made from GMO crops labeled so that conscientious consumers could avoid their purchase. Although there is no evidence that it is dangerous for humans to consume such plants as food or to eat animals that have been raised on GMO crops (see review by Nicolia et al., 2014), it is argued that labeling of such products for consumers who are philosophically opposed to the existence of GMOs and wish to avoid them is no different than requiring labeling of contents of products for sale in the supermarket for the presence of peanuts or gluten or carcinogens or chemical preservatives. Consequently, it is likely that DNA analyses to demonstrate specific genetically engineered genes will be used in the future for identifying GMO plants and that these analyses will be involved in various types of litigation.

1.3 Plant Ecology

Forensic plant ecology is strongly rooted in the techniques of taxonomy but requires a much broader understanding of ecosystems and how they change over time. A single plant species usually is not sufficient to link a victim or a suspect to a specific site unless there is only one location within hundreds of miles where that species is found (for example, the almond Bermuda grass on the golf course in the Bahamas; see Chapter 7). Rather it usually is the unique collection of species that links one to a particular habitat type. Even so, the collection of species identified may only provide a clue to a number of separate sites of similar composition such as a riparian area along a stream or vegetation type associated with a particular elevation or latitude. That may be enough to imply a relationship that could add to the body of evidence in a case against a suspect. For example, the presence of the leaves of a particular kind of spruce tree and the relative absence of another conifer species in Michelle Wallace's hair helped identify the type of habitat where we might expect to find her body (see Chapter 9).

The majority of ecological cases involving forensic botany have been associated with flowering plants (angiosperms) in part because they are the dominant constituents of the flora of most terrestrial ecosystems from arid grasslands to tropical rain forests. Studies have shown that the examination of diatoms (see Chapter 10) and bryophytes (e.g., mosses; see Virtanen et al., 2007) can be extremely useful forensically although clearly they have been underutilized to date.

Stamens are modified leaves and are the sites in flowering plants associated with **microsporogenesis**. In gymnosperms, this process is accomplished by modified leaves called **microsporophyls**. Microsporogenesis leads to the formation of sperm cells within pollen grains. Similarly, **megasporogenesis** leads to egg cell formation inside the developing plant ovary. The analysis of pollen (see Chapter 10) from flowering plants and conifers (gymnosperms) can be very helpful in determining the time of year an event occurred as well as in identifying a unique ecological site. Pollen can indicate the source of drug plants since they may have pollen present that is from native plants where the drug plant sample originated.

When associated with a corpse, pollen can indicate the time of year a death occurred or may provide clues to the actual location of the killing. Even fossil pollen has been used to identify a crime scene (see Chapter 10).

Many palynologists specialize only in pollen analyses. Spore analysis from nonseed plants may call for expertise in one to several subdisciplines within the plant sciences. For example, help may be needed from experts in **pteridology** (ferns), **bryology** (mosses, etc.), **lichenology**, **diatomology**, **algology**, and especially **mycology** (fungi).

Finding experts in spore and pollen identification, especially those with forensic experience, needs to become a standard weapon in the forensic scientist's arsenal. Such plant scientists often are associated with natural history museums and may be located by contacting museum directors. Certain geologists often have experience with pollen or spore identification and are associated with colleges and universities or work in the oil and gas industry.

2. HOW TO BECOME A FORENSIC BOTANIST

There are no full-time positions in forensic laboratories for a forensic botanist at the present time. This has been justified by claiming there is not enough work for a full-time or even a regular part-time position. One of our purposes here is to alert the forensic laboratory industry of the utility of possessing such botanical expertise. There clearly is a need for trained people in forensic plant science, especially at the state and federal levels. Training in forensic botany should include acquisition of general scientific skills, botanical knowledge, and courtroom skills, as well as participation in the activities of professional forensic organizations.

2.1 Scientific and Botanical Training

Our first recommendation for people interested in forensic botany is to obtain graduate degrees in the plant sciences (plant ecology, plant systematics, horticulture, plant pathology, plant morphology and anatomy, palynology, algology, etc.) while obtaining some training in general forensic science. Once gainfully employed as a plant scientist, one can afford to participate on an ad hoc basis in forensic botany. Alternatively, one can seek classical forensic training in a formal program and learn about the appropriate areas of scientific methodology and botany through another college or university program even while working in a forensic laboratory.

2.2 Participation in Professional Forensic Organizations

We strongly urge botanists in North America who are interested in forensics to get involved with forensic societies such as the **American Academy of Forensic Sciences (AAFS)** and/or the **International Association for Identification (IAI)** that hold annual national meetings and sometimes regional meetings on forensic matters. For example, AAFS frequently offers workshops that focus on testimony techniques and other matters that can be very useful (such as evidence handling, crime scene procedures, etc.). These professional forensic organizations, however, do not recognize forensic botany as a separate discipline and the forensic botanist or aspiring forensic botanist must affiliate her/himself with a subsection that seems appropriate. In AAFS, this might be the Pathology/Biology section, for example.

Similarly, members of professional forensic organizations should join existing botanical organizations such as the Botanical Society of America or similar societies in other countries. As members they could encourage these organizations to institute sections that are aimed at individuals interested in forensic plant science.

2.3 Importance of Teamwork in Performing Forensic Analyses

Teamwork is an especially valuable approach in performing forensic analyses. As a rule, we have operated as a team with one of us doing the initial analysis and the other checking his/her results and conclusions. On occasion we have worked with one or two additional botanists. Teamwork helps to discover errors or bring out disagreements in interpretation that then can be discussed and resolved. Sometimes this results in more research or additional consultations. Teamwork is most effective when you and your other team member(s) can work on the evidence separately. If disagreements arise and cannot be reconciled through discussion, you may need to review and/or redo your work until agreement is reached.

Often you and your team member(s) work in isolation from investigators in other disciplines that may be involved in the same case. You simply prepare and submit your report. Sometimes your work is important only in the investigation prior to arrests or trials. Surprisingly, sometimes your forensic data may cause someone to confess their crime. Once caught in a lie, a suspect may recognize the futility of his/her claim of innocence. At other times, your data and conclusions may not be needed. Occasionally you will be asked to testify in court. We have found when serving on the defense team that having meetings where the various forensic disciplines present their findings can be very helpful when you are preparing your testimony for court. When working for the prosecution, you should receive a summary of the evidence against the defendant from the prosecutor. Teamwork may determine the verdict in court.

3. FORENSIC BOTANY IN THE COURTROOM

We like to think of the courtroom as a modified classroom where our job is to educate the jury, the judge, and the lawyers about what we do, how we do it, and what we have learned. Consequently, you must be careful to define your more complicated terms in simple words and analogies so as not to confuse the courtroom audience but instead lead them to an understanding of your conclusions. Some of your jurors may not even have a high school education whereas others may have advanced degrees. You need to educate some without talking down to others. PowerPoint presentations can be extremely informative if the slides are kept as simple as possible with minimal wording (Figure 11.1). Too much information in a visual results in the viewer focusing on reading everything that is on the slide and not really listening to what you are saying.

Judges have great respect for juries, especially when they are sequestered because these people who are isolated from their normal lives have given their time for difficult work in felony cases. Look at the jurors as you give your evidence if you can. You may tell if they understand what you are saying. A good sign is that they are taking notes as you testify. Grand jury members often are allowed to ask you questions as you are giving testimony. Some judges

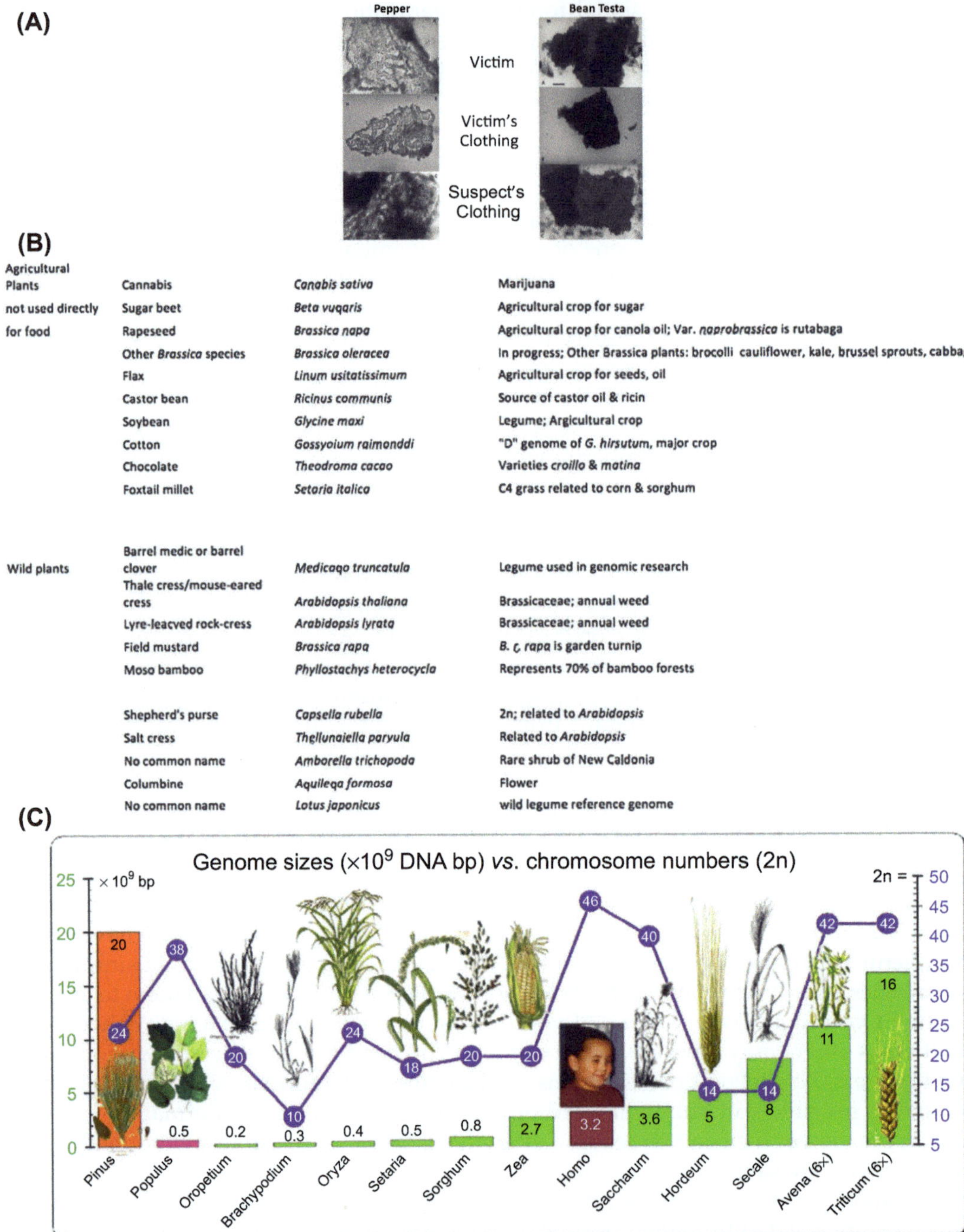

FIGURE 11.1 (A) Representation of a simple slide comparing food items found in fecal samples that link a suspect to a victim. (B) A slide with too much information and print that is too small to read easily. (C) A beautiful and intellectually satisfying illustration that contains too many concepts for the viewer to grasp during a short visual presentation. *Reprinted with permission from Ajmal et al. (2014).*

allow seated jurors in formal trials to do this as well. Think with great care about how you answer questions from jurors or lawyers or the judge because your answers can have a major impact on a verdict.

3.1 Preparation for Testifying in the Courtroom

In performing forensic work in the laboratory or in the field, it is best to know as little about the case as possible so as not to form opinions that might be prejudicial to drawing unbiased conclusions about the data as we discussed in Chapter 2. In this regard, you don't want to know a great deal about the case in advance. However, you of course need to have clear instructions of what you are being asked to do as well as some relevant circumstances related to the evidence you are examining. But you do not want to know what you are expected to find. Similarly, the less you know about the people involved (plaintiff, suspect, victim, etc.), the better. However, once you have completed your analyses, it is most important that you know as much as possible about the case before you are to testify, including other evidence that may relate to your testimony so that you don't say something in ignorance that might damage the case for you side (prosecution or defense). Preparation is the key to effective testimony and you need to be well prepared. You may refer to notes while testifying if you have a lot of data or other details.

3.2 Events in the Courtroom

In USA courtrooms, the only person wearing a black robe over his or her street clothes is the judge, the presiding officer of the proceedings. When the judge enters the courtroom, all present are asked to rise, and then the judge or the bailiff asks most people in the room to be seated. The judge acts as the chairperson of the courtroom proceedings, deciding when jurors and witnesses are called to participate and when they are permitted to enter the courtroom. The judge determines when lawyers for both sides of a case are allowed to speak and can interrupt anyone speaking at any time. It is important that you understand courtroom protocol for the jurisdiction in which you are testifying. In venues outside the USA, for example, judges and attorneys may wear red or black robes, white wigs, hats of all sorts of design, etc.

You usually must wait outside the courtroom before you testify. This keeps you from being influenced by courtroom events that precede your testimony. Waiting to testify may involve a few hours but sometimes is a matter of days. You may find yourself waiting with other witnesses. No matter how tempting, you should not discuss your testimony with others you encounter while waiting.

Additional aspects of courtrooms where you are testifying in criminal proceedings can cause you to feel uneasy because of the nature of the setting. First, there is the presence of the accused and her/his friends as well as associates of the victim. In some cases a large number of interested bystanders also are present. And then, there may be the presence of the media, including reporters, artists, and radio personnel. Television in the courtroom often is a travesty, inviting numerous interruptions to normal legal proceedings. This can reduce major felony cases including death penalty ones to the level of daily TV fare. In the daily TV shows, the judge alone usually decides the civil trial verdict. Such judges are professional actors often with limited judicial experience. In major felony cases that are televised, commentators whose legal backgrounds are less significant than their TV persona will comment on selected televised sections of the trial proceedings.

Avoid getting involved with TV programs or TV production during the duration of a trial and try to ignore the courtroom disruptions made to accommodate the media.

Common advice to courtroom witnesses is to say as little as possible under all circumstances. One common ploy is to ask you a question beyond your expertise. For example a forensic botanist could be asked a question best answered by a mechanical engineer. Your strength here is to say, "I don't know." But sometimes you should explain why you don't know. For example, in a recent case, both the prosecution and the defense had hired forensic botanists. Before either plant scientist had visited the crime scene, the scene and the surrounding area had been scraped bare of all vegetation. The scraped materials were heaped up, inspected, and discarded by federal and local agents who were only searching for bones. This meant many potentially important botanical clues were gone. For example, trampled intact vegetation indicating a path or paths used by those involved in the homicide was lost. Also, ecological clues indicating what sorts of animals had interfered with human remains were destroyed. The forensic scientists had to determine what sort of vegetation surrounded the body of the victim from crime scene investigators' photos, some of which illustrated that the investigators themselves had made changes to the scene. In this situation, a forensic botanist should point out these issues so that the addition of "I don't know" has a meaningful context.

If you are asked a question during cross-examination that you don't understand, ask for clarification or say "I don't understand the question" or "Could you please rephrase the question". Sometimes questions are designed to mislead you and so you must be absolutely clear about what is really being asked. Also consider your answer as an opportunity to interject pertinent information that was not asked for but that may strengthen your testimony or appropriately preface your answer to the question that was asked. Always remember, however, that long-winded answers may impair the attentiveness of your audience. The extremely long and detailed testimony of the DNA expert in the O.J. Simpson trial is a classic example of the latter.

3.3 Deterrents for Academic Plant Scientists to Participate in Legal Investigations

Most academic scientists have been trained extensively in the appropriate scientific methods for conducting studies from the formulation of testable hypotheses to application of appropriate statistical analyses and the proper drawing of conclusions. This is true of those scientists working in both the plant and animal sciences. Forensic laboratory technicians are trained largely to perform techniques and generally have not been rigorously trained as scientists. Although the **National Commission of Forensic Science (NCFS)** has rightly recognized that crime labs should form partnerships with universities to infuse more "science" into the forensic laboratory, academic scientists are reluctant to get involved for the following reasons: academic work loads and the university reward system, concerns about testifying in court, and fears of retaliation.

3.3.1 Academic Work Loads

Academic scientists generally have heavy demands of teaching and research. Teaching involves not only classroom instruction but also training of undergraduate and graduate students to become competently trained laboratory and/or field scientists. Additionally, academics must be constantly working to obtain grants to support their research. Although administrative rewards to faculty for teaching have improved over the past few decades, publication productivity is what brings in the grant money and usually leads to higher rewards within the university (i.e., promotion, tenure, salary, etc.). Grant money also provides overhead costs that help to run the university bureaucracy. Hence, faculty may be reluctant to take on additional projects when they are already committed to more than 40h per week. An appeal to their commitment to public service on behalf of the university might be a useful strategy to get them involved. Public service often is viewed as a positive activity for faculty by university administrations.

3.3.2 *Concerns about the Act of Testifying in Court*

Depositions, formally written summaries in lieu of actual court appearance, are losing favor in the courts because there is no opportunity for the opposition's lawyers to cross-examine the testimony contained in a deposition. Hence, it is considered to favor unjustly one side over the other. Court testimony is a job expectation for workers in forensic laboratories, and they have plenty of colleagues to provide advice. Academic scientists, on the other hand, have had little courtroom experience and may have formulated unrealistic ideas about courtroom operations through movies and television portrayals. Hence, academic botanists are often reluctant to become involved with forensic work because they are concerned that they might have to testify in court. Of course, scientists in academic institutions are accustomed to presenting their research results orally at scientific meetings and fielding questions that test their knowledge and interpretation of their findings. In a way, courtroom testimony and cross-examination is similar to scientific presentations, although in the courtroom the results may alter someone's life significantly.

The opposition lawyers are usually less friendly in court than most of your scientific colleagues and students have been. They may try to discredit you or try to make you look foolish in front of the jury so that your testimony might be questioned or disregarded altogether. Fortunately, there are a number of sources available to ease the anxiety of testifying that can help you with useful pointers for preparation and testimony (e.g., Bowers, 2013; Brodsky, 1999, 2004; Cohen, 2007; MacHovec, 2012; Matson, 1994).

3.3.3 *Concerns about Potential Retaliation*

Academic plant scientists may be somewhat reticent about speaking out publically in criminal court in front of the accused and his/her supporters for fear of retaliation. Historically, and especially in fictional accounts, testifying against criminals has sometimes resulted in retaliation against witnesses. However, expert witnesses generally are not the targets of retaliation. Furthermore, such retaliation is more likely to happen in countries outside of the USA. Only once in more than 30 years has either of us been threatened, and then it was when we were testifying for the defense, not the prosecution. Such threats tend to come over anonymous electronic communications and deserve to be ignored. If you feel truly threatened, notify police or other law enforcement officers. Yes, retaliation is a possibility, but the probability is extremely small.

3.3.4 *Deciding whether or Not to Take a Case*

This advice applies especially to people who are not employed by the justice system, although such employees might benefit from these considerations as well. Listen carefully to the person who approaches you about taking a case. Try to avoid letting them tell you what they want you to "find" or "prove" through your efforts. If you do your work properly, sometimes you cannot satisfy the desires of the requester. If you decide a case may be served by your expertise, you often will need to request documents concerned with the case including autopsy reports; photos taken by crime scene workers; collections of relevant materials including vegetation samples taken from vehicles, homes, clothing, and shoes that are associated with the victims, suspects, and the crime scene.

4. FORENSIC BOTANY IN THE TWENTY-FIRST CENTURY

One of the things the National Academy of Sciences (NAS) Report (NAS, 2009) got right was the need for more "science" in forensic science. A scientist is trained to look at data and formulate a predictive hypothesis. Then the hypothesis is tested experimentally to see if it is supported by additional data. In testing the hypothesis, the scientist must run appropriate controls. Understanding the proper way to establish controls is another essential part of the scientist's training. If the hypothesis is not supported by new data, the hypothesis is rejected or modified. And then it is tested again and again and again. Unfortunately, forensic technicians are not trained this way but are instead taught how to perform certain tests and procedures.

It is clear from the NAS Report and the recommendations of the NCFS (NCFS, 2015; Table 11.2) that there will be an increased effort to establish good scientific procedures for many areas now covered by forensic laboratories, especially in certain areas (e.g., arson, hair, blood spatter, and bite mark analyses). Although DNA analyses have been used as examples of impeccable forensic science, it is now obvious that identifications based on DNA are not as exclusive as once claimed (see Hsu, 2015b). Improved training of forensic technicians and more cooperation between scientists at universities and the law enforcement system will be an essential element to future progress. It is also clear that our inability to control world population growth and the continued degradation of the environment through chemical pollution; physical degradation; and the absence of sustainability practices in agriculture, industry, and personal living will contribute to increased world poverty as this century progresses. And if we have learned one thing, we know that poverty is linked to crime.

4.1 Code of Scientific Responsibility: A Professional Code for Forensic Scientists

Following the 2009 NAS Report, the **Education, Ethics and Terminology Inter-Agency Working Group (EETIWG)** of the National Science and Technology Council's Subcommittee on Forensic Science proposed a code of ethics and scientific responsibility in 2010. The EETWIG reviewed the existing codes of ethics for many professional forensic organizations. Although each code differed considerably among the various disciplines involved, all identified the following central principles: (1) working within professional competence, (2) providing clear and objective testimony, (3) avoiding conflicts of interest, and (4) avoiding bias and influence, either real or perceived.

4.2 The NCFS Recommendations

The NCFS has aired a preliminary document with a number of recommendations. It is anticipated that a formal statement of these recommendations will be available by the time

TABLE 11.2 Proposed National Code of Professional Responsibility for Forensic Science and Forensic Medicine Service Providers[a]

The National Code of Professional Responsibility ("Code") defines a framework for promoting integrity and respect for the scientific process and encouraging a research-based culture. To increase public confidence in the quality of forensic services, each forensic science and forensic medicine service provider must meet the requirements enumerated below:

1. Accurately represent his/her education, training, experience, and areas of expertise.
2. Pursue professional competency through training, proficiency testing, certification, and presentation and publication of research findings.
3. Commit to continuous learning in the forensic disciplines and stay abreast of new findings, equipment, and techniques.
4. Promote validation and incorporation of new technologies, guarding against the use of nonvalid methods in casework and the misapplication of validated methods.
5. Avoid tampering, adulteration, loss, or unnecessary consumption of evidentiary materials.
6. Avoid participation in any case where there are personal, financial, employment-related or other conflicts of interest.
7. Conduct full, fair, and unbiased examinations, leading to independent, impartial, and objective opinions and conclusions.
8. Make and retain full, contemporaneous, clear, and accurate written records of all examinations and tests conducted and conclusions drawn, in sufficient detail to allow meaningful review and assessment by an independent person competent in the field.
9. Base conclusions on generally accepted procedures supported by sufficient data, standards and controls, not on political pressure or other outside influence.
10. Do not render conclusions that are outside one's expertise.
11. Prepare reports in unambiguous terms, clearly distinguishing data from interpretations and opinions, and disclosing all known associated limitations that prevent invalid inferences or mislead the judge or jury.
12. Do not alter reports or other records, or withhold information from reports for strategic or tactical litigation advantage.
13. Present accurate and complete data in reports, oral and written presentations, and testimony based on good scientific practices and validated methods.
14. Communicate honestly and fully, once a report is issued, with all parties (investigators, prosecutors, defense attorneys, and other expert witnesses), unless prohibited by law.
15. Document and notify management or quality assurance personnel of adverse events, such as an unintended mistake or a breach of ethical, legal, scientific standards, or questionable conduct.
16. Ensure reporting, through proper management channels, to all impacted scientific and legal parties of any adverse event that affects a previously issued report or testimony.

[a] *Based on a draft from the National Commission on Forensic Science, National Institute of Standards and Technology, 2015. http://www.regulations.gov/#!docketBrowser;rpp=25;po=0;dct=N%252BFR%252BPR%252BO;D=DOJ-LA-2015-0004.*

this book is published and change may be underway unless implementation is slowed by Congress as has happened in the past with similar attempts to reform the criminal justice system in the USA. Among the recommendations of the NCFS are a code of professional responsibility, some changes in the pretrial discovery process, and establishment of a national training facility.

4.2.1 A Code of Professional Responsibility

Consequent to the recommendations by EETIWIG, the NCFS is recommending a unified code of "professional responsibility" that would govern workers in all forensic disciplines (Table 11.3). The NCFS rejected the term "ethics" as it was considered too broad a topic to adequately define. However, since there is no official organization governing the forensic

TABLE 11.3 Recommendations from the National Commission on Forensic Science Concerning Pretrial Discovery of Forensic Materials[a]

1. When a party proposes to use forensic evidence in a criminal case, the adversary party should be provided with access to the underlying items examined (if reasonably available) and with detailed information about the kinds of analyses conducted and methods used to evaluate those items; the testing conducted on those items; the observations made; the opinions, interpretations and conclusions reached; and the bases for those observations, opinions, interpretations, and conclusions.
2. Access to such information should be made in sufficient time for the adversary party to make effective use of the information.
3. Access to such information should be equally available to both sides, regardless of which side is proposing to use the evidence.
4. Access to such information should be enforceable by the parties through the courts.

[a] *Based on a draft from the National Commission on Forensic Science, National Institute of Standards and Technology, 2015. http://www.regulations.gov/#!docketBrowser;rpp=25;po=0;dct=N%252BFR%252BPR%252BO;D=DOJ-LA-2015-0004.*

work of botanists, forensic plant sciences was not addressed in the initial draft. This omission highlights the need for a professional organization of forensic botanists.

4.2.2 *Pretrial Discovery Recommendations*

The NCFS has made other recommendations to improve fairness in the courtroom. If the recommendations of the NCFS on pretrial discovery are adopted (see Table 11.3), forensic scientists may have to use more creative ways to meet these conditions. For example, the prosecution in a criminal case may have to allow the defense team to have their experts participate directly in the analysis when there is not enough material for both sides to analyze separately such as we proposed in one of our cases (see Chapter 5).

4.2.3 *A Proposed National Training Facility*

The NCFS (2015) has proposed the establishment of a national training facility that would simultaneously provide training in the scientific process and procedures employed by forensic scientists for mixed groups of lawyers, judges, law enforcement personnel, and forensic technicians. This would be modeled after a program that exists in Texas and would include the areas listed in Table 11.4. Notable by their absence in the NCFS recommendations are certain topics such as hair analyses, bite mark analyses, and **graphology** (analysis of the physical characteristics and patterns of handwriting). The lack of scientific rigor and inappropriate testimony related to hair and bite mark analyses has been discussed recently (see Balko, 2015a,b,c,d,e; Hsu, 2015a). Numerous reversals of verdicts involving testimony related to hair analyses by the FBI and bite marks have come to light through the efforts of The Innocence Project (see Chapter 2).

It is unfortunate that the biological training proposed by NCFS would be limited to the analyses of human DNA and would not include nonhuman animal DNA (nuclear and mitochondrial) or plant DNA (nuclear, mitochondrial, and the recent advances using cpDNA). Although botanical analyses as outlined in this book have many valid applications for forensic work, they are not mentioned under the "Pathology/Biology" section in this government publication. Perhaps the plant sciences are ignored because at the time

TABLE 11.4 Areas Proposed Initially for an Interdisciplinary Training Facility[a]

This training facility would simultaneously train lawyers, judges, law enforcement personnel, and scientists

1. Digital multimedia
   a. Speaker recognition
   b. Imaging technologies
   c. Digital evidence
   d. Facial identification
2. Biology/DNA
   a. Human DNA analysis
3. Chemistry/instrumental analysis
   a. Controlled substances
   b. Fire debris and explosives
   c. Geological materials
   d. Materials (trace)
   e. Gunshot residue
   f. Toxicology
4. Physics/pattern
   a. Firearms and tool marks
   b. Footwear and tire tread
   c. Friction ridge
   d. Questioned documents
   e. Blood stain pattern analysis
5. Crime scene/death investigation
   a. Forensic anthropology
   b. Dogs and sensors
   c. Fire scene and explosives
   d. Medical/legal death investigation
   e. Odontology

[a] *Based on a draft from the National Commission on Forensic Science, National Institute of Standards and Technology, 2015. http://www.regulations.gov/#!docketBrowser;rpp=25;po=0;dct=N%252BFR%252BPR%252BO;D=DOJ-LA-2015-0004.*

the report was prepared there was no professional society recognizing contributions by plant scientists to forensic science or providing certification for practitioners of forensic plant science.

5. ESTABLISHMENT OF FORENSIC BOTANY OR FORENSIC PLANT SCIENCE AS A SUBDISCIPLINE WITHIN THE FORENSIC SCIENCES AND BOTANICAL SCIENCE ORGANIZATIONS

Greater utilization of forensic botany and establishment of a professional forensic botany association could be an important asset to the criminal justice system in the twenty-first century. Emphasis should remain on the relatively inexpensive and traditional procedures of plant science investigations although better and less expensive DNA identification procedures should be forthcoming. Because palynology has so much potential for providing forensic evidence, yet is underutilized, there should be a concerted effort made by law enforcement and academics to develop forensic palynology training programs.

Another area that may become important in forensic science is the use of bacterial studies. Scientists at the University of Colorado have found that burial of dead pigs weighing about 20 kg or mice produces predictable changes in soil bacterial communities that may be useful in terms of determining when a body was buried (see Carter et al., 2015; Metcalf et al., 2013).

Future training in forensic plant sciences should include emphasis on traditional approaches and explore those areas that currently are underutilized.

Training of plant scientists should include an awareness of the contributions they can make to forensic analyses. Similarly, training of law enforcement investigative personnel as well as present and future forensic scientists should include basic training in plant sciences.

Currently, there are no training programs in forensic botany, and the subject is rarely mentioned in forensic circles. We have conducted a few forensic botany workshops but this doesn't approach a true training program. The recent NAS Report (2009) does not consider forensic botany as a forensic discipline, and it does not appear in the recent governmental publication on how best to handle and preserve biological forensic evidence (Technical Working Group on Biological Evidence Preservation, 2013).

There are few printed resources that deal with forensic botany for those interested in learning more about it. Introductory forensic classes should include a unit on forensic botany and advanced courses in forensic botany would be most useful. This is hampered by the fact that current forensic textbooks do not have chapters on forensic botany (e.g., Houck and Siegel, 2010; James and Nordby, 2014; Saperstein, 2015). An edited textbook on environmental forensics has only one plant chapter and that deals with dendroecology, useful only in determining the age of trees (Murphy and Morrison, 2015). One textbook that purports to focus on forensic biology (Gunn, 2009) does describe some forensic uses for plants along with a few case examples including some of our work, but this topic needs to be expanded in future editions. Two edited books have been published under the title of "Forensic Botany" (Coyle, 2005; Hall and Byrd, 2012), although their emphases differ considerably from our present volume. One does provide some useful tips on field collection of plant materials (Hall and Byrd, 2012) and both provide introductions into basic plant biology. Additionally, Coyle (2005) provides a good discussion of forensic palynology.

5.1 The Need to Establish a Forensic Plant Science Organization

Accreditation requires establishment of standards of training and responsibility in forensic botany by a knowledgeable group of forensic plant scientists. Presently, there is no such national organization. Establishment of such a group could be accomplished by forming a subgroup for forensic botany within the Botanical Society of America (BSA; botany.org) that already is aware of the important contributions of forensic botany to crime solving (see http://botany.org/PlantTalkingPoints/Crime.php). If you join an appropriate professional scientific organization (for example, the BSA, the Ecological Society of America, or a comparable national organization in other countries), encourage and participate in the establishment of a section for forensic botany. This national organization also could be influential in making the forensic community more aware of the ways to use botanical evidence. Perhaps this will occur in the near future. Until then, forensic botany will remain as an ad hoc forensic activity.

References

Ajmal Ali, M., Gyulai, G., Hidvégi, N., Kerti, B., Al Hemaid, F.M.A., Pandey, A.K., Lee, J., 2014. The changing epitome of species identification – DNA barcoding. Saudi Journal of Biological Sciences 21, 204–231.

Balko, R., February 13, 2015a. How the Flawed 'Science' of Bite Mark Analysis has Sent Innocent People to Jail. The Washington Post. http://www.washingtonpost.com/news/the-watch/wp/2015/02/13/how-the-flawed-science-of-bite-mark-analysis-has-sent-innocent-people-to-jail/.

Balko, R., February 18, 2015b. Attack of the Bite Mark Matchers. The Washington Post. http://www.washingtonpost.com/news/the-watch/wp/2015/02/18/attack-of-the-bite-mark-matchers-2/.

Balko, R., February 17, 2015c. It Literally Started with a Witch Hunt: A History of Bite Mark Evidence. The Washington Post. http://www.washingtonpost.com/news/the-watch/wp/2015/02/17/it-literally-started-with-a-witch-hunt-a-history-of-bite-mark-evidence/.

Balko, R., February 20, 2015d. The Path Forward on Bite Mark Matching — and the Rearview Mirror. The Washington Post. http://www.washingtonpost.com/news/the-watch/wp/2015/02/20/the-path-forward-on-bite-mark-matching-and-the-rearview-mirror/.

Balko, R., April 8, 2015e. A Bite Mark Matching Advocacy Group Just Conducted a Study That Discredits Bite Mark Evidence. The Washington Post. http://www.washingtonpost.com/news/the-watch/wp/2015/04/08/a-bite-mark-matching-advocacy-group-just-conducted-a-study-that-discredits-bite-mark-evidence/.

Bowers, C.M., 2013. Forensic Testimony: Science, Law and Expert Evidence. Academic Press.

Brodsky, S.L., 1999. The Expert Expert Witness. American Psychological Association.

Brodsky, S.L., 2004. Land Plants and DNA Barcodes: Short-Term and Long-Term Goals. Coping with Cross-Examination and Other Pathways to Effective Testimony. American Psychological Association.

Bruni, I., De Mattia, F., Galimberti, A., Galasso, G., Banfi, E., Casiraghi, M., Labra, M., 2010. Identification of poisonous plants by DNA barcoding approach. International Journal of Legal Medicine 124, 595–603.

Carter, D.O., Metcalf, J.L., Alexander Bibat, A., Rob Knight, R., 2015. Seasonal variation of postmortem microbial communities. Forensic Science, Medicine, and Pathology 11, 202–207.

Chase, M.W., Salamin, N., Wilkinson, M., Dunwell, J.M., Kesanakurthi, R.P., Haidar, N., Savolainen, V., 2005. Land plants and DNA barcodes: short-term and long-term goals. Philosophical Transactions of the Royal Society of London Series B, Biological Sciences 360, 1889–1895.

Cohen, K.S., 2007. Expert Witnessing and Scientific Testimony: Surviving in the Courtroom. CRC Press.

Coyle, H.M., 2005. Forensic Botany: Principles and Applications to Criminal Casework. CRC Press.

Ferri, G., Corradini, B., Ferrari, F., Santunione, A.L., Palazzoli, F., Alu', M., 2015. Forensic botany II, DNA barcode for land plants: which markers after the international agreement? Forensic Science International: Genetics 15, 131–136.

Grant, M.C., Mitton, J.B., Linhart, Y.B., 1992. Even larger organisms. Nature 360, 216.

Guiry, M.D., 2012. How many species of algae are there? Journal of Phycology 48, 1057–1063.

Gunn, A., 2009. Essential Forensic Biology, second ed. Wiley.

Hall, D.W., Byrd, J.H., 2012. Forensic Botany: A Practical Guide. Wiley-Blackwell.

Houck, H.M., Siegel, J.A., 2010. Fundamentals of Forensic Science, second ed. Academic Press.

Hsu, S.S., April 18, 2015. FBI Admits Flaws in Hair Analysis over Decades. The Washington Post. http://www.washingtonpost.com/local/crime/fbi-overstated-forensic-hair-matches-in-nearly-all-criminal-trials-for-decades/2015/04/18/39c8d8c6-e515-11e4-b510-962fcfabc310_story.html.

Hsu, S.S., May 29, 2015. FBI Notifies Crime Labs of Errors Used in DNA Match Calculations since 1999. Washington Post. http://www.washingtonpost.com/local/crime/fbi-notifies-crime-labs-of-errors-used-in-dna-match-calculations-since-1999/2015/05/29/f04234fc-0591-11e5-8bda-c7b4e9a8f7ac_story.html.

James, S.H., Nordby, J.J., 2014. Forensic Science: An Introduction to Scientific and Investigative Techniques, fourth ed. CRC Press.

MacHovec, F.J., 2012. The Expert Witness Survival Manual. iUniverse self publishing. www.iuniverse.com.

Matson, J.V., 1994. Effective Expert Witnessing. CRC Press.

Metcalf, J., Wegener-Parfrey, L., Gonzalez, A., Lauber, C.L., Knights, D., Ackermann, G., et al., 2013. A microbial clock provides an accurate estimate of the postmortem interval in a mouse model system. eLife 2, e01104.

Murphy, B.L., Morrison, R.D., 2015. Introduction to Environmental Forensics, third ed. Academic Press.

NAS, 2009. Strengthening Forensic Science in the United States: A Path Forward. National Research Council of the National Academies, National Academies Press, Washington, DC.

NCFS, 2015. National Commission on Forensic Science, National Institute of Standards and Technology Draft Statement. http://www.regulations.gov/#!docketBrowser;rpp=25;po=0;dct=N%252BFR%252BPR%252BO;D=DOJ-LA-2015-0004.

Nicolia, A., Manzo, A., Veronesi, F., Rosellini, D., 2014. An overview of the last 10 years of genetically engineered crop safety research. Critical Reviews in Biotechnology 34, 77–88.

Piperno, D.R., 2006. Phytoliths. A Comprehensive Guide for Archeologists and Paleoecologists. AltaMira Press.

Saferstein, R., 2015. Forensic Science: From Crime Scene to the Crime Lab, third ed. Prentice Hall.

Technical Working Group on Biological Evidence Preservation, 2013. The Biological Evidence Handbook: Best Practices for Evidence Handlers. National Institute of Standards and Technology, U.S. Dept. of Commerce. http://dx.doi.org/10.6028/NIST.IR.7928.

Virtanen, V., Korpelainen, H., Kostamo, K., 2007. Forensic botany: usability of bryophyte material in forensic studies. Forensic Science International 172, 161–163.

Appendix I

Materials Needed for Plant Anatomy Analyses

1. Compound microscope with a trinocular head (two eyepieces and a phototube) equipped with 4×, 10×, 25× or 30×, and 40× objectives (Appendix VI Figure). The most useful ones for plant cells are 10× and 25×. An oil immersion objective is not necessary unless you want to examine pollen.
2. A digital camera that can be connected to a laptop or desktop computer is essential for making a photographic record of observations. If you plan to purchase a new microscope, get one that includes a digital camera and software. You also can purchase a digital camera and software that can be used on virtually any microscope with a phototube (Figure I.1).
3. Dissecting binocular microscope with phototube for looking at seeds or other large, opaque objects.
4. Glass microscope slides (30 × 100 mm). These can be purchased with a label space, with a frosted end for labeling or plain (Figure I.2(A)). The plain glass slides are the least expensive. They can be labeled with the use of a diamond marking pen. We recommend

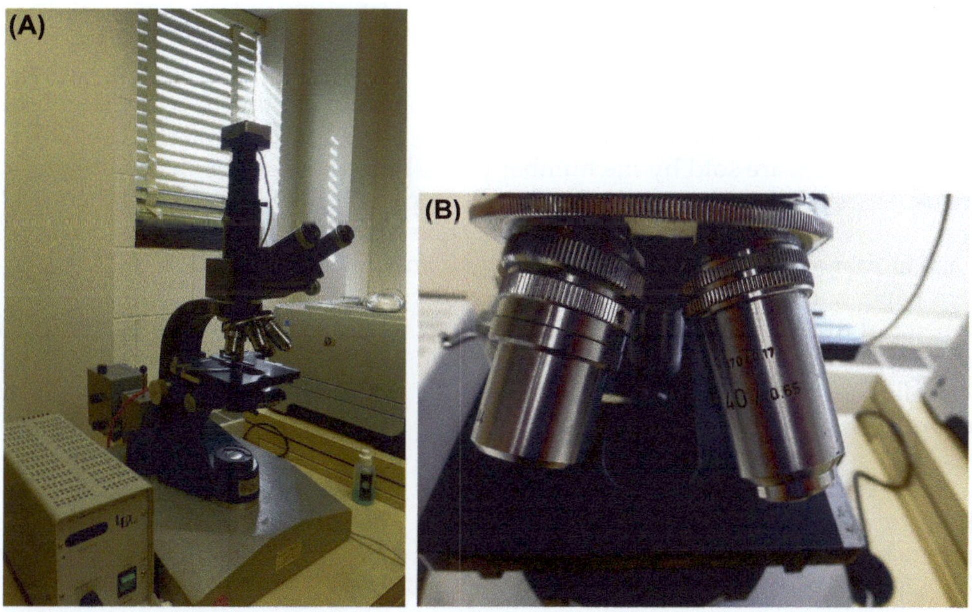

FIGURE I.1 (A) A simple binocular compound microscope equipped with a phototube and digital camera that connects to a computer. (B) A 40× objective.

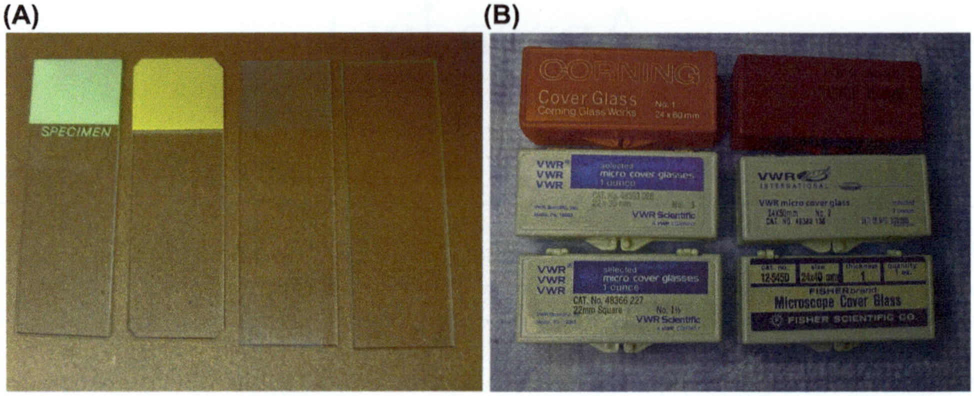

FIGURE I.2 There are a variety of glass microscope slide types (A) The three slides to the left can be labeled with pencil. The plain slide at the right is least expensive but requires a diamond marker to etch a label on the slide. The colored slide ends are more expensive and come in a variety of colors useful for keeping track of a source, for example. The frosted slide (third from the left) is a good compromise. (B) Coverslips (cover glasses) come in a variety of sizes and thicknesses. We use mostly 22 × 22 and 22 × 30 coverslip sizes at No. 1 thickness. 22 or 24 × 60 covers most of a standard microscope glass slide.

the frosted type for ease of recording information directly on the slide as they are less expensive than the colored ones. It is best to use a sharp pencil to label slides because the label will not dissolve in any of the solvents used. The ink of some marking pens is soluble in ethanol and/or in the clearing agents. Clear tape can be placed over the pencil label of the final pencil label on the slide to prevent smearing or accidental erasure of the label.

5. Glass coverslips (cover glasses) can be purchased in a range of sizes and thicknesses (No. 1 is the thinnest). We use No.1 thickness coverslips as they provide the best resolution. The most common sizes we use are 22 × 22 and 22 × 30 (Figure I.2(B)). Although slides are sold by the number (usually ½ gross per box), coverslips are usually sold by the ounce. At least one company (SPI Structures/Probe Inc. http://www.2spi.com) thinks this is absurd and sells coverslips in packets of 100.
6. Small mortar and pestle for gentle grinding of known plant samples.
7. Lens paper and lens cleaner for cleaning microscope lenses.
8. Eppendorf tubes with snap-tops (1, 2, and 3 mL) or glass vials with tight screw caps for sample storage.
9. The following tools:
   - a. Disposable pipettes
   - b. Dissecting needles (straight ones preferred)
   - c. Small forceps
   - d. Single-edged razor blades
   - e. Fine scissors
10. Kimwipes or other absorbent paper (never use these on microscope lenses).

11. Chemicals (see Appendix IV)
    a. Ethanol (95% and 100%)
    b. Formalin (100%)
    c. Clearing agent
    d. Mounting medium (e.g., Permount®)
    e. Plant cell wall stains

Some suggested sources for microscope slides and materials.

Carolina biological supply	http://www.carolina.com
Ted Pella, Inc.	http://www.tedpella.com
VWR	https://us.vwr.com

Appendix II

Techniques for Slide Preparation to Examine Gastrointestinal Specimens or Fresh/Frozen/Cooked Food Samples with the Compound Light Microscope

1. PROCEDURE FOR PREPARING GASTROINTESTINAL SAMPLES

Samples of human gastrointestinal contents should be disinfected prior to examination. Regardless of the treatment prior to receiving a sample, we routinely add enough 100% formalin to all liquid samples to yield a final concentration of 5–10% formalin. These samples typically will have to be diluted even further with 10% formalin prior to analysis. If a sample is frozen, we thaw it in 10% formalin. Dry samples are hydrated in 10% formalin. This ensures that any infectious agents present are neutralized.

2. PROCEDURE FOR PREPARING WET MOUNTS

Concentrated samples may need to be diluted with 10% formalin prior to analysis. Larger items (visible clumps, seeds, etc.) should be removed and saved for separate examination. After removal of larger items, the remaining sample should be thoroughly mixed so that a representative subsample can be removed and diluted for microscopic analysis. Additional subsamples can be examined to verify that the sample was well mixed.

Add one or more drops of gastric, intestinal, vomitus, or fecal material suspended in water or 10% formalin (or other fixative) directly on a prelabeled glass microscope slide placed on a flat, level surface. Dried materials should be hydrated prior to examination and may need to be dispersed into smaller fragments. This can often be done on the surface of the slide by separating clumps with clean dissecting needles. Select an appropriately sized coverslip for the size of the drop and touch the coverslip to the solution at one end of the slide (Figure II.1). Gently lower the coverslip slowly so as to push the fluid toward the other end of the slide. Do this slowly so as not to create any air bubbles. A dissecting needle is a good device to insert under the free end of the coverslip and use to lower the coverslip. Once the coverslip is down, then slowly withdraw the needle from under the coverslip. You may need to gently push on the coverslip with the needle or forceps to flatten the material. Excess water on the slide surface should be absorbed with a paper towel. Observe immediately as such preparations will soon dry out. Photograph all items you find and keep notes as to the relative abundance of each object. This is especially important when comparing fecal samples. (Note: if preparation is too dense to allow you to resolve individual items, you may need to further dilute a subsample from the source sample and make a new slide.)

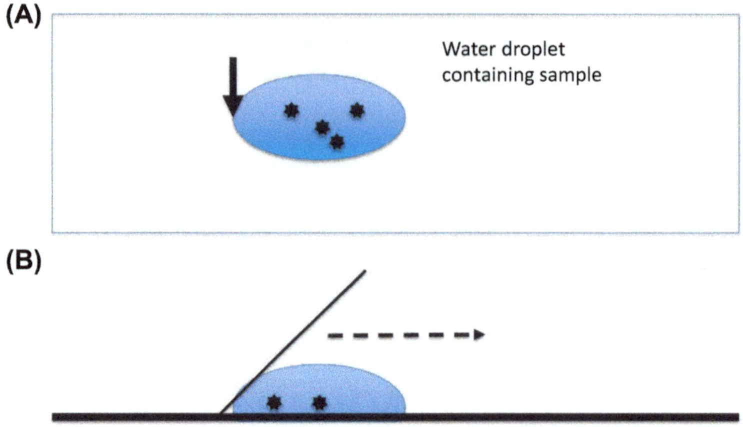

FIGURE II.1 (A) Looking down on a glass microscope slide with a drop of water containing the sample. Arrow indicates where to place the edge of coverslip. (B) Side view. Position the coverslip at an angle toward the other end of the slide and lower slowly with a dissecting needle in the direction of the arrow. Gently slide the tip of the needle from under the coverslip.

3. PROCEDURE FOR PREPARING PERMANENT MOUNTS

This procedure will allow you to prepare permanent slides of your known plants and materials from gastrointestinal tract samples for long-term storage and/or for others to examine at a later time. You may wish to do this procedure under a fume hood or in a well-ventilated space due to the volatility of the alcohols, clearing agents, and the mounting medium. Aqueous mounting media are available commercially but do not provide the permanence for long-term storage. The procedure is summarized in Table II.1.

3.1 Preparing Permanent Slides

1. Add one or more drops of gastric, intestinal, vomitus, or fecal material suspended in 10% formalin (or 70% ethanol) directly on a prelabeled glass microscope slide placed on a flat surface.
2. Allow the fresh subsample to dry completely in air (this may take several hours so you might want to employ a warming tray at 40–50 °C to speed this up). Drying attaches the plant material to the glass surface. Commercial pretreated slides with an adhesive are available but probably are not necessary if the slides are not moved at this stage.
3. Although the slide appears to be dry, there still may be residual water trapped within the plant cells. In order to make a clear, permanent preparation, it is important to remove all of the water by next exposing the material to increasing concentrations of ethyl alcohol (ethanol; 70, 95, 100%). Treat the slide as described below.
4. To remove solutions from the slide, hold the labeled end and tip the slide so that the solution will run off the opposite end into an appropriate waste container (follow local guidelines for dealing with chemical wastes). Carefully flood the slide with the next

TABLE II.1 Summary of Procedure for Preparing Permanent Slides

Steps	Time	Rationale
1. Add one or a few drops of gut material suspended in 10% formalin directly on a prelabeled glass microscope slide placed on a flat surface. (Stain if preferred.)		
2. Dry to affix suspension to glass.	Variable	Adheres plant slide by drying to slide surface.
3. Flood dry slide with 70% ethanol.	5 min	Begin removal of residual water.
4. Replace 70% with 95% ethanol.	3–5 min	Continue dehydration.
5. Replace 95% with 100% ethanol.	3–5 min or until dry	Complete dehydration.
6. Replace 100% with Histo-Clear®.	3–5 min	Eliminate alcohol which will not mix with the mounting medium.
7. Add standard mounting medium (e.g., Permount®, Fisher Scientific) and add coverslip.		
8. Allow slides to harden while lying flat for at least 24 h before observing with microscope. Excess mounting meeting can be removed at this time from the slide or coverslip with a paper towel soaked with clearing agent.		

Note: Since we often preserve plant material in 70% ethanol, we simply dry the sample completely on the slide and go directly to the 100% ethanol step. Although this causes extensive shrinkage of animal cells, it does not affect plant cell walls although the cytoplasm will shrink.

solution in the series using a pipette or eye dropper. Be careful not to disturb materials on the slide. Alternatively, the slides can be dried thoroughly on a warming plate for several hours, thus eliminating the need for the higher percentage alcohols.

5. The alcohol must be replaced with two changes of a clearing agent such as Histo-Clear® that replaces the alcohol which will not mix with the permanent mounting medium. Before the Histo-Clear® evaporates, add the permanent mounting medium (we recommend Permount® which if properly applied will last for many years). The mounting medium will mix with the clearing agent but not with alcohol or water. Add just enough mounting medium to fill the area under the size of coverslip you are going to use. Excess mounting medium will require much longer drying times and medium may ooze out and fuse the slide to the slide storage container. Next, carefully apply the coverslip as described for the wet mount.
   Note: There are aqueous mounting media available that avoid the drying procedure but in our experience those slides don't store as well as Permount® mounted slides.
6. Freshly prepared permanent slides should be placed on a flat surface and be allowed to harden 24–48h or more before attempting to observe them under a microscope.[1] Excess mounting medium that may have adhered to the back of the slide or the top of the coverslip should be carefully removed using a tissue soaked in the clearing agent prior to allowing the slides to harden. The use of the warming tray can be helpful here. Care should be taken in handling the freshly made slides as the coverslip can be moved or removed rather easily before the mounting medium has hardened. After hardening any movement of the coverslip may damage the coverslip as well as the plant material.
   The mounting medium should be allowed at least a week to harden before placing the slide in a slide box for storage. If the mounting medium is still soft, it may flow into the slide box and harden, forming a permanent bond between the slide and the slide box. Slide boxes can be obtained from a scientific supply house.

4. STAINING PLANT MATERIALS PRIOR TO OBSERVATION

We often observe the plant materials or gastrointestinal contents directly without using dyes. Staining of the plant cells prior to observation can increase the contrast and produce better photomicrographs. Alternatively, the coverslip can be removed after examining the fresh mount, the stain applied, and the coverslip replaced.

We suggest the use of safranin O (red) for staining plant cell walls. Safranin is especially good for staining cellulose. Staining of fecal samples with safranin can also help identify which of the debris present are plant fragments. Safranin will also stain your hands so it is good to wear gloves when working with this dye. Another dye, toluidine blue, stains some of the additional components associated with cell walls (e.g., lignins) and may also be useful.

[1] We maintain what we call a "slop scope" for quick examinations of slides. The slop scope is an old microscope that we don't care about and may not have very good optical properties, but it allows us a low-power (10×) look at the newly made permanent slides to get an idea of what is there. Sometimes we eager science types just can't wait.

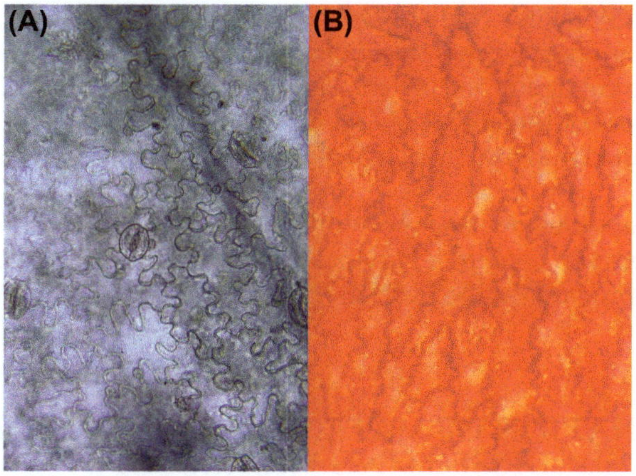

FIGURE II.2 Unstained preparation of lettuce leaf epidermis (A) compared with a leaf somewhat overstained with safranin (B). *Photomicrographs by author.*

Safranin is prepared in alcohol and diluted with water so that it can be applied directly to the material already in water on the slide or 70% ethanol (e.g., fresh mounts). One must experiment a bit to determine the right dilution of the stock dye (see Appendix IV) to use as overstaining can be a problem (Figure II.2). We usually start with the working solution of safranin and dilute it further depending on the intensity of staining. Staining is accomplished almost immediately. It is best to practice staining on known materials to determine a dilution that works best for you. If making a permanent slide, add the dye to the sample on the slide and then follow steps 2–6 above.

5. STAINING PREVIOUSLY UNSTAINED FRESH MOUNTS AND CONVERTING FRESH MOUNTS TO PERMANENT SLIDES

It is possible to stain an unstained fresh mount or make a permanent slide of a fresh mount by carefully removing the coverslip and following the procedure for staining and/or making permanent slides. Care must be taken to remove all the material that usually adheres to the coverslip and add it back to the slide. Also, material may be moved around on the slide to distribute it more evenly, keeping in mind the area that ultimately will be covered by the size of the coverslip you plan to use. Surprisingly, this harsh treatment will not affect your ability to identify the same materials in the permanent slide.

6. EXAMINATION OF SLIDES WITH A COMPOUND MICROSCOPE

Examination of fresh mounts or permanent mounts that have hardened generally requires only the use of 4×, 10×, and 25× objectives. The "high dry" 40× objective is seldom necessary due to the much larger size of plant cells as compared to most animal cells. Fresh-mounted slides should have their backs wiped dry before use. Permanent mounts should be examined only with the 10× or 25× taking care not to touch the objective to the slide. If any mounting medium adheres to the stage or the objective, it should be immediately washed off with a small amount of clearing agent and dried. However, be aware that the clearing agent can penetrate into the lenses of the objective and damage them. One often can increase contrast with stained or unstained material by placing colored filters between the light source and the stage.

Preferably, the compound microscope will be equipped with a phototube and a digital camera so that a permanent copy of all images can be stored on your computer. Inexpensive digital cameras and software can be purchased and used on any microscope equipped with a phototube.

7. EXAMINATION OF MATERIALS WITH A BINOCULAR DISSECTING MICROSCOPE

Opaque and/or larger items that are too large to view with a compound microscope but difficult to see with the naked eye can be observed with the aid of a dissecting microscope. It is best to use a dissecting microscope with a phototube so that you can prepare a permanent record with a digital camera. A small ruler can be photographed with the specimen to verify size (see also Appendix III). Similarly, a good handheld digital camera with close-up capabilities can be used for this purpose.

Appendix III

Making Accurate Measurements with the Microscope

There are several ways to obtain size measurements on cells or other objects viewed with the compound trinocular microscope (binocular compound microscope with a phototube for a digital camera). Typically, all measurements are made in millimeter (mm) or microns (μm). The most accurate method is to calibrate an ocular micrometer (reticle) for each microscope objective using a stage micrometer. One simply purchases a microscope slide with an embedded linear scale called a stage micrometer from a scientific supply house (see ahead). The stage micrometer is placed on the microscope stage as you would any slide. An ocular micrometer or reticle is placed within one of the eyepieces. If these items were not included when your microscope was purchased, they can usually be obtained from the microscope manufacturer or from a scientific supply house. The microscope manufacturer can explain what size reticle to purchase for the ocular (eyepiece) you have, and how to insert the reticle into the appropriate eyepiece. The procedure for doing the calibration can be found online (see examples below). However, if you probably wish to indicate size information on the final digital or hardcopy picture, we recommend the following procedure for measurements.

Photograph a stage micrometer using each microscope objective and compare that directly with the photograph of the object (Figure III.1). Different digital cameras may add different magnification depending on manufacturing parameters as well as the length of the phototube on the photomicroscope. Remember that the camera is replacing the magnification of the eyepiece objectives usually with a lower magnification. Hence a photomicrograph taken while the photographer is viewing the object with a total magnification of 250× (a 25× microscope objective times the 10× eyepiece), what the camera sees might be only a total magnification of 100× (e.g., 25× times 4×). Since printing of photomicrographs often involves additional increases in magnification, this approach is especially useful for providing reasonably accurate measurements relative to the final reproduction. One simply prints the digital picture of the stage micrometer using a given microscope objective (e.g., 25×) along with the print of the object taken with that same microscope objective (Figure III.2).

Dissecting (stereo) microscopes also can be calibrated using the photograph technique. Simply photograph a ruler with a millimeter scale using the various magnifications available. The scale can then be superimposed on the image (Figure III.3). Editing software or even slide-making software can be used to superimpose a measurement bar directly on a photomicrograph.

Note: If more than one microscope is being used, each should be calibrated for measurements since there may be differences in the magnifications of the available microscope objectives, the length of the phototube used with the camera, etc.

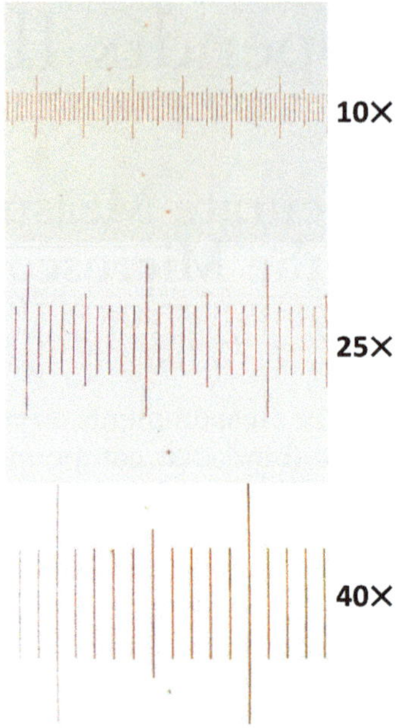

FIGURE III.1 Stage micrometer viewed with three compound microscope objectives (10×, 25×, and 40×). Each unit is 10 μm. The long bars are 0.1 mm apart (100 μm).

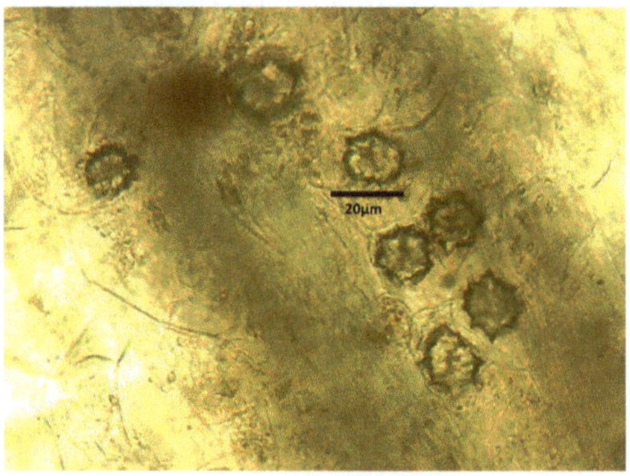

FIGURE III.2 Druse crystals from okra photographed at 40×. *Photomicrograph by author.*

FIGURE III.3 Raspberry seed and section of millimeter scale photographed separately and superimposed at the same magnification. The seed is approximately 2.5 mm in length. *Photographed by author using a stereomicroscope with a 10× objective.*

Commercial source for micrometers
http://www.microscope-depot.com/stagemicro.asp.
http://www.microscope-depot.com/reticles.asp.

Calibration techniques using an ocular and a stage micrometer
http://www.mecanusa.com/microscope/micrometer/micrometer-use_en.html.
http://homepages.gac.edu/~cellab/chpts/chpt1/ex1-3.html.
http://labprotocolsonline.blogspot.com/2011/06/measurements-ocular-and-stage.html.

Appendix IV

Composition of Solutions Used in Preparation of Plant Cells for Microscopic Examination

1. POTASSIUM IODIDE–IODINE SOLUTION (LUGOL'S SOLUTION)

Lugol's solution is a mixture of iodine and potassium iodide (KI) in distilled water that can be used for staining starch in plant cells (see Chapter 4, Figure 4.17). We prepare a stock solution and dilute it as needed. Commercial preparations are available although some of them are designed for use as dietary supplements. If you don't have access to the necessary reagents to make Lugol's, we suggest this commercial source: http://www.hometrainingtools.com/iodine-solution-lugols-30ml/p/CH-IODINE/.

Iodine–potassium iodide solution (1.5%). Dissolve the KI first and then add the iodine crystals as that will help the iodine crystals dissolve.

Iodine crystals	0.3 g
Potassium iodide (KI)	1.5 g
Distilled water	100.0 mL

2. SAFRANIN O SOLUTION

Stock Solution	
Safranin O	2.5 g
95% ethanol	100.0 mL
Working Solution	

Dilute 10 mL of stock solution with 90 mL distilled water.

Note that further dilution of the working solution may be necessary depending on the material to be stained. Commercial solutions are also available.

3. TOLUIDINE BLUE SOLUTION

Commercial preparations of a 1% solution of toluidine blue are available.

4. ETHANOL SOLUTIONS

Ethanol (ethyl alcohol) is the preferred alcohol to be used for preservation of tissues and for processing of samples for making permanent slides. It can be purchased at 100% or 95%

concentrations. Chemicals are used to remove the water in 95% ethanol to make 100%. Thus, it is important to keep 100% ethanol in tightly sealed containers as water vapor can dilute the alcohol. Because 95% is considerably less expensive than 100%, you should dilute 95% to yield lower percentage solutions. A simple technique is to take the number of mL of 95% ethanol equal to the final percentage you wish to make and dilute to 95. For example, 70 mL of 95% ethanol diluted to 95 mL by the addition of 25 mL of water provides a solution very close to 70% ethanol. Likewise, 700 mL diluted to 950 mL yields almost a liter of 70% ethanol.

5. CLEARING AGENTS

Clearing agents are used to replace alcohols and any remaining water from tissues prior to addition of the mounting medium when making permanent slides (see Appendix II). They also are necessary for preparing blocks of tissues to be embedded in wax for routine sectioning with a microtome prior to making permanent slides. The best clearing agent is xylene but because of its high toxicity it has been replaced with a number of commercial products that are much less toxic (some claim no toxicity) to lab workers and much less flammable than xylene. We routinely use Histo-Clear® but you can find other brands available online.

5.1. Sources for Histological Clearing Agents

https://www.nationaldiagnostics.com/histology/product/histo-clear.
http://www.hemo-de.com/.
http://www.carlroth.com/website/en-com/pdf/RotiHistol_E.pdf.
http://www.bio-world.com/productinfo/2_33_238/2892/Bioclear-tissue-clearing-agent.html.

6. FORMALIN SOLUTIONS

Formalin is a solution of formaldehyde gas in water. 100% formalin is a solution of 37% formaldehyde. Commercial preparations of 100% formalin can be diluted with water (tap water is sufficient) to make a 10% formalin working solution. Alternatively, commercial preparations of 10% neutral buffered formalin (NBF) can be purchased from many chemical supply companies. NBF is formalin to which additional chemicals have been added to reduce the acidity of the solution. You can make your own NBF inexpensively from 100% formalin:

10% Neutral buffered formalin	
Sodium phosphate, monobasic	4.0 g
Sodium phosphate, dibasic	6.5 g
Formaldehyde, 37% (100% formalin)	100 mL
Distilled or tap water	900 mL

Appendix V

Methods for Verification of Feces and Vomitus

1. FECES

Verification of feces can be done by detection of the bile pigment urobilinogen in the suspected fecal material using the Edelman's reagent. A dilute solution of urobilinogen can be purchased to use as a control.

http://hazards.tees.ac.uk/Rams3/general/DETAILScoshh1.cfm?recordID=2753.

http://www.astm.org/DIGITAL_LIBRARY/JOURNALS/FORENSIC/PAGES/JFS11489J.htm.

2. VOMITUS

A quick screening test for the presence of gastric fluid can be used to identify stains which are suspected to be vomit. Known vomit samples as small as 0.5 mm in diameter have tested positive using this procedure. A small portion of the suspected stain is removed to a black porcelain spot plate along with a dried vomit control. Several drops of whole cow's milk are pipetted into each well. The spot plate is placed into a humidity chamber at 38 °C approximately for 30 min. The spot plate wells are then examined visually and with a stereobinocular microscope. The occurrence of coagulation or curdling indicates gastric enzymes are present in the sample (see Figure 6.8).

Reprinted with permission from Schneck, W.M., 2004. Cereal murder in Spokane. In: Houck, M.M. (Ed.) Trace Evidence Analysis: More Cases in Mute Witnesses, Academic Press, pp. 165–190.

Appendix VI

Maceration of Wood Samples for Microscopic Examination

The following procedure was developed by Dr William (Ned) Friedman.

Day 1

Wood fragments were prepared by finely slicing wood from each tree species with a sterile razor blade. These fragments were transferred into labeled 2-mL Eppendorf tubes. One milliliter of a solution of 1:4:5 hydrogen peroxide:deionized (DI) water:glacial acetic acid was added to each tube. The samples were then placed in a 60 °C oven for 48–60 h. When first inserted into the oven, the caps are left off and the tubes are covered loosely with foil. This prevents gas build-up and allows the tubes to adjust to the new temperature. After an hour, monitoring periodically, the foil is removed with the caps firmly placed onto each tube. This procedure dissolves the middle lamella, which serves as the glue that holds cells together.

Day 3

The solution in each tube is replaced with DI water (1.5 mL) without disturbing the cells at the bottom and left for 1 h followed by a second change of DI water (1.5 mL). This procedure washes the cells of any remaining acetic acid or hydrogen peroxide.

Day 4

A third and final change of DI water (0.75 mL) is performed. After 1 h, 0.75 mL of 100% ethanol is added to the existing DI solution and left for 1 h. This solution is then replaced with 1.5 mL of 100% ethanol and left for 1 h. A final change is performed replacing the ethanol with 1.5 mL of 1% safranin O in 100% ethanol. The tubes are then placed in a 60 °C oven for 20 h. The same protocol for introduction of the tubes to the oven treatment is used as on the first day.

Day 5

Do two 5-min rinses of 1.5 mL 100% ethanol and replace with a 1.5 mL 1:1 ethanol:Hemo-De® solution for 2 min. Subsequently, the solution is rinsed by addition of 1.5 mL Hemo-De®, once for 2 min and again for 5 min. Hemo-De® is a clearing agent similar to Histo-Clear® (see Appendix IV). Next, the samples are sonicated for approximately 5 s to separate the tissues permanently. This process should be followed with careful examination of each sample, as some may need more sonication. However, oversonication should be avoided as this may result in cell destruction.

Samples can be mounted on slides using Permount® for later examination with a compound microscope (see Appendix II for mounting procedure). Cells can be identified as to type and/or measured to characterize a particular species.

Appendix VII

Photographic Atlas Contents

Common name	Species name	Plant family
Anise	*Pimpinella anisum* L.	Apiaceae
Apple	*Malus* Mill.	Rosaceae
Apricot	*Prunus armeniaca* L.	Rosaceae
Artichoke	*Cynara scolymus* L.	Asteraceae
Artichoke, Jerusalem	*Helianthus tuberosus* L.	Asteraceae
Arugula	*Eruca vesicaria* (L.) Cav.	Brassicaceae
Asparagus	*Asparagus officinalis* L.	Liliaceae
Avocado	*Persea americanum* Mill.	Lauraceae
Banana	*Musa acuminata* Colla	Musaceae
Bean, black	*Phaseolus vulgaris* L.	Fabaceae
Bean, garbanzo (chickpea)	*Cicer arietinum* L.	Fabaceae
Bean, green	*Phaseolus vulgaris* L.	Fabaceae
Bean, kidney	*Phaseolus vulgaris* L.	Fabaceae
Bean, lima	*Phaseolus lunatus* L.	Fabaceae
Bean, lupin (garden vetch)	*Vicia sativa* L.	Fabaceae
Bean, pinto	*Phaseolus vulgaris* L.	Fabaceae
Beet	*Beta vulgaris* L.	Chenopodiaceae
Blackberry	*Rubus* L.	Rosaceae
Blueberry	*Vaccinium* L.	Ericaceae
Broccoli	*Brassica oleracea* L. var. *Italica*	Brassicaceae
Cabbage, red and white	*Brassica oleracea* L. var. *capitata*	Brassicaceae
Cantelope	*Cucumis melo* L.	Cucurbitaceae
Caraway	*Carum carvi* L.	Apiaceae
Carrot	*Daucus* L. (spp.)	Apiaceae
Cauliflower	*Brassica oleracea* L. var. *botrytis*	Brassicaceae
Celery	*Apium graveolens* L.	Apiaceae
Cherry	*Prunus cerasus* L.	Rosaceae

Common name	Species name	Plant family
Chives	*Allium schoenoprasum* L.	Liliaceae
Cloudberry	*Rubus chamaemorus* L.	Rosaceae
Coriander (cilantro)	*Coriandrum sativum* L.	Apiaceae
Corn	*Zea mays* L.	Poaceae
Cranberry	*Vaccinium macrocarpon* Aiton	Ericaceae
Cucumber	*Cucumis sativus* L.	Cucurbitaceae
Cucumber, African horned	*Cucumis metuliferus* E. Mey. ex Naud.	Cucurbitaceae
Cumin	*Cuminum cyminum* L.	Apiaceae
Dandelion	e.g., *Taraxacum officianale* L.	Asteraceae
Dill	*Anethum graveolens*	Apiaceae
Eggplant	*Solanum melongena* L.	Solanaceae
Fennel	*Foeniculum vulgare* L.	Apiaceae
Fig	*Ficus carica* L.	Moraceae
Flax	*Linum usitatissimum* L.	Linaceae
Garlic	*Allium sativum* L.	Liliaceae
Grape	*Vitus* L.	Vitaceae
Grapefruit	*Citrus paradisi* L. (hybrid origin)	Rutaceae
Honeydew	*Cucumis melo* L.	Cucurbitaceae
Kale	*Brassica oleracea* L. var. *acephala*	Brassicaceae
Kiwi	*Actinidia chinensis* Planch.	Actinidiaceae
Leek	*Allium porrum* L.	Liliaceae
Lemon	*Citrus limon* (L.) Burm (hybrid origin)	Rutaceae
Lentil	*Lens culinaris* Medik.	Fabaceae
Lettuce	*Lactuca* L. (many varieties)	Asteraceae
Mustard	*Brassica napus* L. var. *napus* (many species)	Brassicaceae
Okra	*Abelmoschus esculentus* (L.) Moench	Malvaceae
Olive	*Olea europaea* L.	Oleaceae
Onion, green	*Allium cepa* L.	Liliaceae
Onion (bulb, many colors)	*Allium cepa* L.	Liliaceae
Oregano	*Origanum vulgare* L.	Lamiaceae
Papaya	*Carica papaya* L.	Caricaceae
Parsley	*Petroselinum crispum* (Mill.) Fuss	Apiaceae

Common name	**Species name**	**Plant family**
Peach	*Prunus persica* (L.) Batsch	Rosaceae
Pear	*Pyrus communis* L.	Rosaceae
Pear, Asian	*Pyrus ussuriensis* Maxim.	Rosaceae
Pea, garden	*Pisum sativum* L.	Fabaceae
Pea, snow	*Pisum sativum* L.	Fabaceae
Pepper, bell and cayenne	*Capsicum annuum* L.	Solanaceae
Pimento	*Capsicum annuum* L.	Solanaceae
Pineapple	*Ananas comosus* (L.) Merr.	Bromeliaceae
Plum	*Prunus domestica* L.	Rosaceae
Pomegranate	*Punica granatum* L.	Lythraceae
Poppy	*Papaver* L.	Papaveraceae
Potato, white	*Solanum tuberosum* L.	Solanaceae
Potato, sweet	*Ipomoea batatas* (L.) Lam.	Convolvulaceae
Raspberry	*Rubus coronarius* (Sims) Sweet	Rosaceae
Rosemary	*Rosmarinus officinalis* L.	Lamiaceae
Oregano	*Origanum vulgare* L.	Lamiaceae
Sesame	*Sesamum orientale* L.	Pedaliaceae
Spinach	*Spinacia oleracea* L.	Chenopodiaceae
Squash, zucchini	*Cucurbita pepo* L.	Cucurbitaceae
Squash, crookneck	*Cucurbita moschata* Duchesne	Cucurbitaceae
Star fruit	*Averrhoa carambola* L.	Oxalidaceae
Strawberry	*Fragaria chiloensis* (L.) Mill.	Rosaceae
Tarragon	*Artemisia dracunculus* L.	Asteraceae
Tomato	*Solanum lycopersicum* L.	Solanaceae
Turnip	*Brassica rapa* L.	Brassicaceae
Water chestnut	*Trapa natans* L.	Lythraceae

Subject Index

Note: Page numbers followed by "f", "t" and "b" indicate figures, tables and boxes respectively.

Vernacular Names Index

Note: Page numbers followed by "f" and "t" indicate figures and tables respectively.

Scientific Names Index

Note: Page numbers followed by "f" and "t" indicate figures and tables respectively.

Kolibri 360 GmbH
Ferdinand-Jühlke-Straße 7
99095 Erfurt, Deutschland
produktsicherheit@kolibri360.de